ECONOMIC MODELS OF MATERIAL-PRODUCT CHAINS FOR ENVIRONMENTAL POLICY ANALYSIS

ECO-EFFICIENCY IN INDUSTRY AND SCIENCE

VOLUME 4

The titles published in this series are listed at the end of this volume.

Economic Models of Material-Product Chains for Environmental Policy Analysis

by

Patricia P.A.A.H. Kandelaars
Department of Spatial Economics,
Amsterdam, The Netherlands

KLUWER ACADEMIC PUBLISHERS
DORDRECHT / BOSTON / LONDON

A C.I.P. Catalogue record for this book is available from the Library of Congress.

ISBN 978-90-481-5255-1

Published by Kluwer Academic Publishers,
P.O. Box 17, 3300 AA Dordrecht, The Netherlands.

Sold and distributed in North, Central and South America
by Kluwer Academic Publishers,
101 Philip Drive, Norwell, MA 02061, U.S.A.

In all other countries, sold and distributed
by Kluwer Academic Publishers,
P.O. Box 322, 3300 AH Dordrecht, The Netherlands.

Printed on acid-free paper

CONTENTS

Preface

This monograph contributes to the quest for sustainability and environmental quality from an environmental economic perspective. It examines the physical and economic aspects of material and product flows, and the policies and strategies designed to reduce related resource depletion and environmental pollution. Various theoretical and applied models are presented that explicitly include physical dimensions. The resulting analyses are economically consistent and physically feasible.

From 1994 to 1998 I have been working on this monograph at the department of Spatial Economics of the Vrije Universiteit in Amsterdam. I have enjoyed joint research with Hans Opschoor, Jeroen van den Bergh, Rob Dellink, Monique Jansen, Fred Lambert and Jan van Dam. Various chapters are based on work with these colleagues.

I am indebted to the Dutch Organization of Scientific Research (NWO) for financial support that allowed me to perform this research at the Vrije Universiteit in Amsterdam.

I would like to acknowledge the consent of various publishers to use material that has appeared previous in journals.

Patricia Kandelaars

Amsterdam, March 1999

CHAPTER 1

INTRODUCTION

1.1. Motivation and approach

Many important environmental problems can be traced back to the use of materials and energy. These link environmental problems of scarcity and pollution to resource extraction and waste emissions. Environmental economics has mainly focused on a partial analysis of environmental problems, as illustrated by separate branches like resource economics', dealing with depletion issues, and 'pollution economics', addressing pollution externalities. As a result, environmental economics tends to neglect the interdependence of environmental problems related to the particular economic stages which occur between extraction of resources and pollution of the environment. Without taking into account the linkages between the separate activities between extraction and waste treatment, the indirect effects of policies may be overlooked. For instance, a reduction in the use of one material to reach a certain level for environmental indicator X may require less use of a particular product, but this may trigger an increase in the use of another product providing the same service but made of another material, and then environmental indicator Y may be negatively affected. Therefore, for policy making it may be important to consider these trade-offs explicitly.

Physical aspects of problems related to physical flows are studied by natural and environmental scientists. However, their studies usually do not consider the economic and behaviourial mechanisms underlying material flows. In policy design, physical or technological possibilities, and economic aspects and behaviour really need to be combined. For example, substitution between products can only take place when it is technically possible and when the substitute product is not too expensive or otherwise not attractive.

This study tries to integrate elements of these different areas. It presents an approach that takes the interactions between depletion and pollution into account and regards the economy as being composed of various stages between extraction and emissions. This will involve linking the economic and physical aspects of the use of materials. More in particular, this study aims to contribute to integrated model-based analyses of resource and pollution problems for policy making. The approach here is based on the concept 'material-product (M-P) chain'. An M-P chain can be defined as a set of linked flows of materials and products so as to fulfil a certain service (Opschoor, 1994). An analysis of an M-P chain can be defined broadly as an analysis of an economic structure of connected material and product flows. In this study an 'economic analysis of an M-P chain' is performed in which allocation and economic processes are studied. In this analysis the interactions between demand, production, recycling and waste management on the level of flows of materials and products are studied. The economic modelling of M-P chains means combining the elements of physical flow and economic models.

On the basis of an economic analysis of M-P chains, it is possible to perform 'chain management' that takes into account the linkages between various economic activities. Other concepts have been proposed to consider linked economic processes in order to study environmental problems, such as 'industrial metabolism' (Ayres, 1989; Ayres and Simonis, 1994) and 'industrial ecology' (Graedel and Allenby, 1995) (see Chapter 2). These are, however, less oriented towards incorporating economic market and behaviourial mechanisms.

Two alternative approaches to examine environmental problems and policies may be distinguished: one based on economic welfare theory, the other a more pragmatic, multidimensional approach. Neoclassical economic theory dealing with the design of environmental issues is based on the concept of environmental externalities, and studies policies that optimally reduce or control those externalities. This approach may be characterized as optimizing. The concept of an 'externality' will be defined more precisely below. For now it suffices to consider them as unwanted and unpriced environmental consequences of economic activities (Löfgren, 1995). Externality-based theory may be adopted as the starting-point for studying the environmental impact of material and product flows, and in particular to obtain insight into the relative and absolute welfare effects of various policies oriented at materials and products.

As an alternative to the neoclassical welfare/externality approach, a multidimensional approach to environmental policy analysis aims to evaluate the physical, chemical, ecological and economic impacts in various dimensions, instead of aggregating all impacts into a single welfare index. The impact of a policy on material and product flows is examined via indicators, such as 'amount of materials', and ecological indices, such as 'acidification' and 'costs of recycling', instead of reducing everything to abstract notions of external (social) costs and benefits. This may be described as a satisfying approach.

This chapter is structured as follows. Section 1.2 describes the concept of M-P chains in more detail. The problem definition and the research questions guiding this study are discussed in Section 1.3. A final section presents the structure of this book.

1.2. Economic Analysis of Material-Product (M-P) Chains

An M-P chain refers to a network of economic activities between extraction and waste treatment, connected via flows of materials and products. An M-P chain is connected to the environment by extraction and waste treatment activities. The basis of an M-P chain is the need or desire of consumers for a service or application. Various products may meet the demand for that service. Their production requires materials and energy. Various economic activities are part of an M-P chain: extraction of materials, production, consumption, recycling and waste treatment. All these activities are connected by physical and monetary flows. The physical flows may be divided into traditional physical flows, measured in kilograms, or into more aggregate environmental indicators, which are based on physical units, such as depletion and acidification units. An M-P chain may include recycling of materials, reuse of products and substitution between materials or products.

Using the concept of an M-P chain various analyses can be performed. A broad definition of 'M-P chain analysis' encompasses both economic and environmental analyses of an economic structure of connected material and product flows. Life-cycle assessment (LCA) is an environmental analysis of an M-P chain, because it examines the environmental impact of a product and its material flows (see Section 4.5). However, an M-P chain is not the basis of a material flow analysis (MFA) or materials accounting because an MFA does not include products explicitly (see Section 4.3).

In this study a more limited definition of M-P chain analysis is used: 'Economic analysis of M-P chains' studies the allocation and economic processes of an M-P chain. In this study the term 'M-P chain analysis' is also used for the narrow definition of an 'economic analysis of M-P chains'. M-P chain analysis allows the study of, for instance, optimization, market equilibrium, market processes, production functions, policy analysis, substitution at different levels, explicit modelling of economic processes, and endogenous behaviour of agents.

Figure 1.1 shows an M-P chain for milk packaging. For illustrative purposes only some of the material and product flows are presented. This example attempts to clarify the various flows of materials and products needed to satisfy the demand for a certain service. Milk can be packed in new or reused glass bottles or in carton packs. New and reused glass bottles are perfectly substitutable. The demand for milk packaging will be met by both glass bottles and carton packs. Production of new glass bottles requires new or recycled glass. Production of carton packs requires cardboard and plastic. The cardboard and plastic waste, originating from discarded carton packs, must be disposed of. The waste glass bottles will either be reused, recycled or dumped.

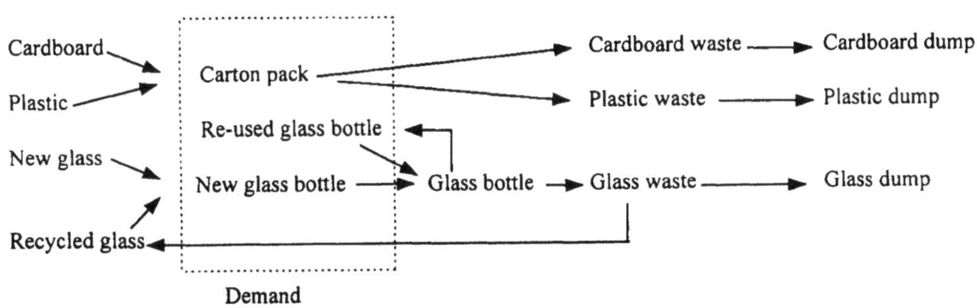

Figure 1.1. An M-P chain for milk packaging.

Thus, a broad definition of M-P chain analysis includes LCA, but in the narrow definition used in this study LCA studies are not included, because they do not include allocation or other economic processes. In the applications in Part III of this study it will be clear that LCA is limited compared with M-P chain analysis, because LCA does not include: (i) an economic optimization function on a social or chain level (see Chapters 6); (ii) market processes and behaviour (see Chapters 7 and 10); and, (iii) scenarios for environmental development (see Chapters 8 and 9). Economic analysis of M-P chains does include at least one of these.

Ideally, the focus of M-P chain analysis is on all relevant environmental aspects. This means that all feasible alternative materials, technologies and products may be taken into account. However, this goes beyond what is practically possible in analytical and data terms. Hence, instead of using 'complete' M-P chains, for the purpose of this analysis these are usually 'truncated', i.e. an M-P chain is reduced to the more relevant parts of it, given the nature of the issues the analysis helps to address (Opschoor, 1994). In other words, a reduction of a complete M-P chain is based on economic, physical and environmental aspects, and on data availability.

M-P chain analysis can provide insight into the flows of various materials and products, their interactions, and the impact of implementation of chain policies. This makes it possible to use models of M-P chains for analysis and sometimes even for predicting the effects of management and public policies, technological development and changes in demand for products or materials.

1.3. Objective of the Study

Research on material flows in economic systems has hitherto mainly focused on describing physical flows in a certain period and region, or related to a particular product (see Chapter 4). Little attention has been devoted to the economic aspects of physical flows. The present study attempts to fill this gap between environmental science, on the one hand, and economics, on the other. The goal is to examine the physical and economic mechanisms related to flows of materials and products, and the possible (policy) scenarios to reduce the environmental and economic problems related to these flows.

The two main objectives of the study are the following.

(1) The formulation of formal alternative economic models of M-P chains in such a way that they enable integrated analysis of the economic and physical impacts of policies.
(2) The empirical application of the models formulated in (1) to various material-product chains and policy instruments.

The above-mentioned objectives will be realized in five interlinked steps.

(A) Considering the basic concepts of physical flows in environmental and economic systems (Chapter 2).
(B) Examining approaches, strategies and policies relevant for M-P chains (Chapter 3).

(C) Reviewing physical and economic models for studying materials or product flows (Chapters 4 and 5).
(D) Formulating economic models to analyse the economic and physical impact of policies in a framework of M-P chains (Chapters 6 to 10).
(E) Applying the models formulated in (D) in empirical case-studies (Chapters 6,8,9 and 10).

1.4. Outline of the Study

In the light of the previously stated research questions, the study is organized into 11 chapters (see Figure 1.2). In Part I, Chapters 2 and 3 constitute the introductory part of the study. Chapter 2 describes some concepts of physical flows in natural systems and relates those concepts to economic systems. The most important ones described are the laws of thermodynamics and the material balance principle. Chapter 3 concludes the introductory part of the study by presenting an overview of strategies and policies that may be used for the management of M-P chains.

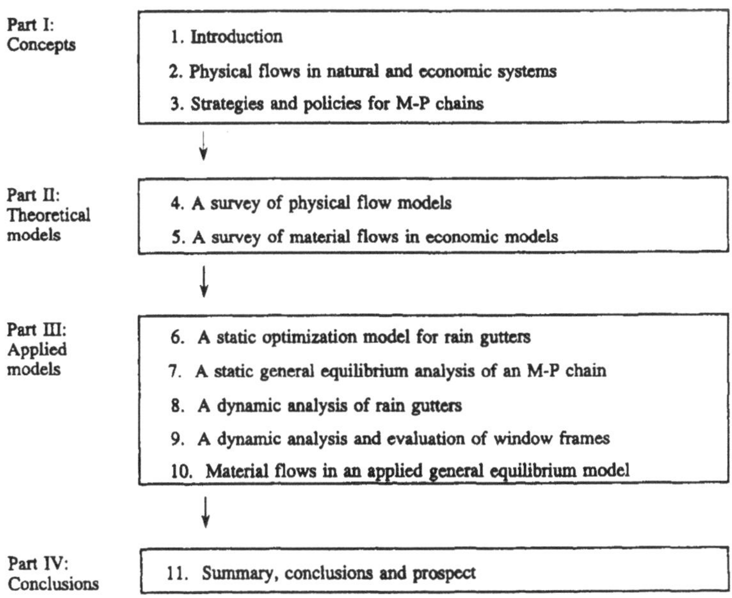

Figure 1.2. Structure of the study.

Models and methods for studying physical flows in economic systems are presented in Chapters 4 and 5 of Part II. Chapter 4 discusses physical models that are often used to describe material flows for a region or a product. Also M-P chain analysis is described in detail. Chapter 5 examines how physical flows have been or can be incorporated in economic models. Chapters 4 and 5 form the basis of the models applied in Part III of the study. These applied models combine the elements of both the physical models of Chapter 4 and the economic models of Chapter 5.

Chapters 6 to 10 of Part III present applications of the models. In Chapter 6 a static optimization model for several M-P chains is discussed. The model is applied to rain gutters to see the effects of various policies. Chapter 7 introduces a general equilibrium model in which the externalities caused by depletion and pollution are optimized by imposing taxes and subsidies. The model may be regarded as representing a general M-P chain of various economic activities that are linked to each other by physical and monetary flows. Chapters 8 and 9 present dynamic models to examine the effects of various policy scenarios on M-P chains. In Chapter 8 a model is applied to rain gutters to explore the economic and physical effects of certain policies. Chapter 9 applies a model to window frames and assesses the impact of policies on physical and economic indicators, and also on environmental indicators, such as acidification and global warming. In Chapter 10 an applied general equilibrium model and a physical input-output model are integrated to analyse the socio-economic effects of material and product policies.

The study concludes with Chapter 11, which presents an overview of the various model-based studies, draws general conclusions and provides some suggestions for further research.

CHAPTER 2

PHYSICAL FLOWS IN NATURAL AND ECONOMIC SYSTEMS

2.1. Introduction

In this chapter the relationship between the economic system, the natural environment and physical flows is discussed. Since 'the economy is embedded in a larger biogeochemical system by material and energy flows', studies of the economy may benefit from a better understanding of the natural environment and its processes (Ayres, 1998). Physical laws such as the material balance (MB) principle apply equally to natural and economic systems. For studying material and product flows in a material-product (M-P) chain, the MB principle may be used to keep track of the inputs and outputs in every stage of the chain.

The organization of this chapter is as follows. Section 2.2 briefly reviews some concepts and insights from the natural sciences relevant for the study of material flows. These include ecosystems, nutrient or material cycles, metabolism and evolution. Section 2.3 discusses similar concepts applicable to economic systems, such as industrial metabolism, material cycles and economic evolution. The laws of thermodynamics and the MB principle are discussed in Section 2.4. Conclusions are presented in Section 2.5.

2.2. Ecosystems, Material Cycles and Evolution

This section describes various basic concepts in natural or ecological systems, i.e. material cycles, metabolism, succession and evolution. These concepts can be helpful when trying to understand the character and systematics of material flows in economic systems. Insights obtained from ecology and evolutionary biology may be useful in M-P chain analysis, which emphasizes physical flows through economic systems.

'Ecology' is the science that studies the relationships of living organisms with their biotic and abiotic (non-living) environment. The abiotic, non-living, parts are the physical and chemical components of an ecosystem, such as climate, sunlight, air and nutrient supplies. The biotic part consists of living organisms, such as bacteria, plants and animals. Ecosystems can maintain their structure due to a continuous input of solar energy.

The relationship between organisms in an ecosystem may be illustrated by a 'food chain', which represents a series of organisms, each feeding on the preceding one. Organisms in food chains 'eat or are eaten'. For a totally closed system this would mean in economic terms that the input of one organism has to be the output of another one (Boulding, 1978). Organisms are a consumer, a producer or a decomposer. 'Producers' such as plants and algae support the life on earth through photosynthesis. They are called 'autotrophs' (Greek for self-feeding) because they provide their own food generally via photosynthesis. These autotrophs are in turn the food for consumers, such as herbivores. Therefore, these consumers are also called 'heterotrophs' (Greek for other-feeding). Decomposers, mainly bacteria and fungi, break down dead organic material and convert it into carbon dioxide, water and nutrients (Odum, 1971). Producers, consumers and decomposers are connected in food chains and webs. A food web is an interconnected network of food chains that includes all feeding relationships in an ecosystem (Chiras, 1994). An M-P chain in an economic system can be considered as similar to a food chain or food web in an ecosystem (see Section 2.3).

Material cycles

Energy and material flows between organisms and the environment are crucial for the continuity and development of ecosystems. "In a totally closed ecosystem the only input is energy from the sun. All the other materials are recycled biologically, in the sense that each species' waste products are 'food' for another species" (Ayres and Ayres, 1996, p. 278-279). Materials or nutrients in ecosystems flow in cycles, the so-called 'biogeochemical' or 'nutrient cycles'. The four most important ones are the nitrogen, carbon, sulphur and phosphorus cycles (Odum, 1971). The carbon and the nitrogen cycle are described shortly by way of illustration.

The basic mechanism of the carbon cycle is that it transforms carbon dioxide, released by living organisms into oxygen and glucose. This is mainly done by photosynthesis. Oxygen is used for respiration and transformed into carbon dioxide. Using sunlight, photosynthesis transforms carbon dioxide (6 CO_2) and water (6 H_2O) into glucose ($C_6H_{12}O_6$) and oxygen (6 O_2). Respiration transforms glucose and oxygen into carbon dioxide, water and energy. Therefore, carbon flows from the abiotic environment (water and air) to the organisms that are part of the food chain, and from there back to the environment. This results in the cycling of carbon in the natural system (Odum, 1971).

Ayres and Ayres (1996) describe the nitrogen cycle with the linkages between the environment and the economy. They especially pay attention to human interference in the natural nutrient cycles which may cause problems for the environment and for organisms including humans. Nitrogen is an essential element in the production of proteins. For economic purposes it is mainly used as a fertilizer (converted into ammonium) in agriculture. When nitrogen comes in contact with the air, nitrogen oxides (NO_x) are created. These are an important factor behind acid rain and acidification of soils and surface water (Alcamo *et al.*, 1990; Van der Voet, 1996).

As opposed to what happens in natural systems, in economic systems materials do not completely cycle, because recycling activities are limited. Full recycling is not technically feasible due to unavoidable leakages. As a result, there is a continuous flow of inputs (extraction) from, and outputs (waste generation) to, the environment.

Metabolism

'Metabolism' is a biological term that refers to 'the totality of internal processes -

both physical and chemical - that supply the energy and nutrients required by an organism as the conditions of life itself' (Ayres and Simonis, 1994). On an aggregate level metabolism may be regarded as the input (intake) of energy and materials into biological cycles where several transformations take place, ultimately ending in waste being generated (the output). Ayres (1989) applies this concept to economic systems referring to 'industrial metabolism', which is discussed in more detail in Section 2.3.

Succession

Changes in natural systems occur on many levels and over short and long periods. Ecosystems change via a process of 'succession'. 'Primary succession' is the sequential development of a community of species where none existed before. 'Secondary succession' is the sequential development of a community after it has been partially or completely destroyed by natural or human causes (Chiras, 1994). At the beginning of secondary succession there are a few types of species with a simple food chain, while in a developed stage of succession more complex ecosystems and food chains may result. Succession is a process that starts with changes or disturbances in the environment. Traditionally, succession is seen as an irreversible process that moves over a unique optimal path until it reaches a unique optimal, stable state. This is called a 'climax ecosystem' that may stay in equilibrium for a long time (Boulding, 1978). This traditional view needs to be revised for several reasons (Holling *et al.*, 1995). One of the reasons is that a disturbance can result in different 'climax' states, depending on the non-unique path towards an equilibrium. Hence, there may be several climax states, or in other words, several equilibria.

Evolution

Changes or development may occur not only by succession but also by evolution. Biological evolution may provide useful insights for understanding technological changes and economic development. Biological evolution can be defined as 'a long-term process that leads to structural, functional, and behaviourial changes in species, known as adaptations' (Chiras, 1994). In biological evolution the selection of species occurs by means of genetic variations and natural selection. Genetic variation is due to differences in the genetic composition of species. It may occur by sexual reproduction in which the young receive a new and unique combination of parental genes, or by mutations caused, for example, by radiation or chemicals in the environment, or it may occur just spontaneously. The Darwinian concept of natural selection refers to a process in which useful variations are preserved and species adapt to changes in their environment. It should be noted that evolutionary biology is broader than Darwinian evolution. Darwin has already stated that 'natural selection has been the main but not the exclusive means of modification' (Gould, 1997).

This short introduction to ecology and evolution may help the reader to understand and conceptualize physical flows, and changes therein, in economic systems via M-P chains, as will be explained in the remainder of this chapter.

2.3. Industrial Metabolism, M-P Chains and Economic Evolution

This section describes in more detail economic analogies of the biological concepts and processes discussed in the previous section. However, these concepts may also be regarded as more than just analogies from the natural environment applied to economic systems, because economic systems are themselves essentially subsystems of the global environmental systems, in particular the biosphere. The concepts addressed in this section form the basis of studying M-P chains.

Industrial metabolism

'Industrial metabolism' is the concept of biological metabolism applied to the economy (Ayres, 1989). Industrial metabolism may be defined as the set of physico-chemical transformations that convert raw materials (biomass, fuels, minerals, metals) into manufactured products and structures and wastes (Ayres and Simonis, 1994). This means that the flows of materials from extraction through the economy and back to the environment are described and examined. This description may be used at various levels (Ayres, 1994). First, on a global level industrial metabolism is 'the whole integrated collection of physical processes that converts raw materials and energy, plus labour, into finished products' (Ayres, 1994, p. 3). Second, on a national, regional or firm level the flows of materials may be measured within a certain geographical region (or, in the case of a firm, locality). Third, and most important for the description of specific material flows is the life cycle of individual materials (or nutrients). For studying the extraction, use and dissipation of materials in a certain region a description of these material flows is needed. Some of the methods to describe the material flows are discussed in Chapter 4. The goal of the industrial metabolism approach is to modify the use of materials to reduce the generation of waste by applying lessons from the natural world (Duchin, 1992).

The term 'economic metabolism' is a more appropriate term than 'industrial metabolism'. The metabolism referred to here is that of the whole economy, not just of industries in a strict sense nor of industrial transformations of physical flows.

A concept which is similar to industrial metabolism is 'industrial ecology'. Socolow *et al.* (1994, p. xviii) explain the difference between these concepts as follows: the word 'industrial' in 'industrial metabolism' refers to the whole society, while in 'industrial ecology' it refers to the activities of specific industrial producers. Industrial metabolism considers transportation in general while industrial ecology considers the automobile or aircraft industry. Another definition is given by Frosch and Gallopoulos (1990), who relate industrial ecology to 'biological ecology' and state that industrial systems should adopt biological notions such as that of a food chain (see Section 2.2). According to Duchin (1992) and Graedel and Allenby (1995), industrial ecology may be interpreted as industrial metabolism extended with a human perspective, in which, for example, consumption and preferences are included. In practice though, industrial ecology and industrial metabolism may be treated as one and the same concept.

M-P chains

An M-P chain in an economic system is similar to a food chain in an ecosystem. The flows of materials may be regarded as analogous to the flows of nutrients and other substances, and the flows of products as analogous to molecules composed of those nutrients within the food chain. The main difference is that the economic system is never closed, while a natural system may be closed on a materials level. An open economic system or open M-P chain means that not all materials (and products) are recycled (or reused) (Ayres and Ayres, 1996).

Economic evolution

Next, the biological concepts of natural selection and evolution may be used to study economic phenomena. Economic 'natural selection' may be interpreted as, for example, the survival of certain firms. Thus, the economy as a whole may change by natural selection (Nelson and Winter, 1982). Besides the Darwinian idea of natural selection, the Lamarckian concept of 'deliberate' or 'goal-oriented' processes in evolution may be relevant for describing economic change.

In economic systems, succession may be considered as a structural change in the sectoral structure of an economy due to changes or developments in technology. Some changes in economic systems may be seen as evolutionary changes, for instance, technological, product or institutional innovations. In the context of M-P chains, succession and evolutionary changes may be regarded as the development of an M-P chain, e.g. the extension or replacement of materials or products.

In contrast with ecosystems, economic succession and evolution have not led to a 'closed' economic system in which all waste products are recycled. Over time an increased need has developed for the natural system to deliver more and more resource inputs and to receive an increasing volume of waste outputs. Thus, economic evolution has gradually given rise to a less 'closed' economy, with an increase of throughput of materials in the economy. Throughput is the flow of materials beginning with raw material inputs, followed by their conversion into commodities, and finally into waste outputs (Daly, 1996). However, the use of several materials does not only proceed straight from the environment through the economy and back to the environment, because of recycling and reuse. The evolution of economic systems is often interpreted as an increase in welfare, while in ecological terms it would be more appropriate to see it as a change in the efficiency of using and recycling materials.

In order to reduce the throughput of energy and materials in economic systems it is useful to apply the ideas of biological evolution to economics. For instance, Boulding (1978) distinguishes *genotypes* and *phenotypes* as follows. Production is the process by which the genotype directs energy and transforms materials into a phenotype (a product). Thus, the genotype, which may be interpreted as knowledge, is one of the three factors of production besides materials and energy.

The traits of organizations may be regarded as economic genotypes. For organizations it is easier to maintain themselves in a constant than in a changing environment and it is also easier to have the same type of organizations than totally different ones (Nelson and Winter, 1982, pp. 9-10). Nelson and Winter call the characteristics of firms 'routines' which are regarded as similar to genes in biological evolution. Although the genes are the same, the behaviour of the firm may be different depending on the environment. According to Faber and Proops (1992), economic genotypes include legal, social and economic institutions, and economic phenotypes cover: the techniques of production that are used; the quantities, types and prices of capital and consumption goods; and the market structure. To change the flows of materials in an economy both economic phenotypes and genotypes may be altered.

2.4. Thermodynamics and the Material Balance Principle

When studying material flows in economic systems evidently thermodynamics should be taken into account. The first and second laws of thermodynamics provide insight and limit the possibilities of materials and energy use that are often not considered in traditional economic modelling. An example is the limited substitution of energy or materials in production that in economic models may be considered unlimited.

The first law of thermodynamics, i.e. the law of conservation of energy, states that in a closed system energy cannot be created nor destroyed, but can only be transformed, for example, from chemical to mechanical or kinetic energy (gasoline transformed to movement). In other words, energy is conserved. Under normal earthly conditions, matter and energy can be regarded as independent categories, which implies that matter (also referred to as 'mass') and energy can be considered as independently conserved (Van den Bergh, 1996).[1] At the level of the earth and the atmosphere, the relationship between the environment and the economy is that of an open system because there is an inflow of energy from the sun. For materials it is a closed system in which there are no inputs from and no outputs to the outside. In an open system there may be inflows and outflows of materials and energy. A third type is an isolated system in which there are no inflows or outflows of either energy or materials.

Under practical conditions the first law may also be applied to materials. This derived law is referred to as the 'law of conservation of matter' or 'the material balance', and states that matter remains preserved. Materials can be transformed or transported within the closed system, but they cannot be made or destroyed. This is also known as the material balance (MB) principle. The implication of this principle is that the materials extracted eventually will be returned to the environment, possibly in an altered, degraded form. The MB principle may be useful to complete or check data sets for consistency, to estimate waste residual output, to reconstruct historical data of emissions, or to account or model material cycles (Ayres, 1989; Ayres and Ayres, 1994 and 1996). For an open system the MB principle implies that

---

[1] The relationship between energy and matter (or mass) is reflected by the famous Einstein equation, $E=mc^2$, stating that energy and matter are equivalent, which implies that the sum of energy and matter is conserved. However, energy and matter will only transformed into each other under very special circumstances, which can be disregarded for the purpose of this study.

all materials that go into a system either accumulate or leave the system.

The alteration or degradation of materials as mentioned in the discussion on the first law of thermodynamics is the basis of the second law of thermodynamics, also known as the 'entropy law'. This second law considers the quality of energy and how it changes. It states that entropy - a measure of unavailable energy - will not decrease in an isolated system (Ruth, 1993). There are several ways to interpret this second law (Peet, 1992). One is that when energy is used, it is degraded. This means that energy is transformed into a less concentrated form. For example, the concentrated energy of gasoline is transformed into movement and heat, which is a less concentrated form of energy. Thus, energy cannot be 100% recycled because some of the energy is always lost into heat as unavoidable dissipation. Another way to explain this law is that the amount of energy available for work, the 'exergy', is reduced when it is used (Ayres, 1978; Ruth, 1993). For example, gasoline contains more potential energy or work than the heat which results from its combustion. Boulding (1978) states that the entropy law has a time-arrow, i.e. is a process that has a direction in time. This time-arrow implies that entropy increases over time, i.e. it is an irreversible change. The law of evolution states that 'complexity increases in terms of differentiation and structure' (Boulding, 1978, p. 10).

The second law may be applied analogously to materials, implying that the quality of materials decreases when they are used. This may be referred to as the 'fourth law of thermodynamics'. This may be interpreted as indicating that materials cannot be recycled infinitely or completely. However, this notion of finite or incomplete recycling of materials has been debated. Bianciardi *et al.* (1993) argue that 'complete recycling is physically possible if a sufficient amount of energy is available', but they add that this amount of energy is so great that its production function will be unsustainable. Furthermore, recycling is also limited by economic reasons, such as the lack of markets or the high costs of recycling.

Although all economic systems necessarily behave in accordance with the laws of thermodynamics, the interfaces between economics and thermodynamics are rarely considered or are even neglected (Georgescu-Roegen, 1971; Faber *et al.*, 1987; Perrings, 1987; Ayres, 1994; and, for an overview, see Amir, 1994). Some studies discuss the use or the consequences of the laws of thermodynamics in environmental economics (Ruth, 1993, 1995a and 1995b; Dung, 1992).

Energy and material flows form the basis of economic and environmental systems. In an M-P chain, the environment provides the economy with raw materials and energy, which are transformed into consumer products. After consumption the materials and energy are discarded in the environment. In the context of M-P chains the MB principle is especially important. It allows one to determine the input and output in every stage of the chain. In M-P chain analysis, the MB conditions are imposed on all materials and products that are considered. The 'entropy law' for materials can be included in an M-P chain analysis to account for a decrease in the quality or the incomplete recycling of materials as constraints in the examination of M-P chains.

2.5. Conclusions

This chapter has introduced several concepts that are essential for studying physical flows in environmental and economic systems. It was argued that the concepts of industrial metabolism, material flows and M-P chains in economic systems are analogous to the biological concepts of metabolism, nutrient flows and ecosystems, and useful in the study of material flows in economic systems. The laws of thermodynamics and the derived MB principle also apply to the physical dimensions of economic systems. The first law applied to materials, via the MB principle, may be used to keep track of the material flow through the economy at different stages of the M-P chain. From the second law of thermodynamics it may be concluded that complete recycling is not possible, or at least very costly, due to the degradation of materials. The framework of material flows, the MB principle and limited recycling forms the basis for examining M-P chains. For example, an M-P chain consistent with the MB principle may facilitate an accurate analysis of the opportunities for closing or changing the material and product flows through the economy.

In addition to static concepts, dynamic concepts like succession and evolution were discussed. While natural evolution has given rise to complete recycling of materials, evolution in economics has not yet led to complete cycling of materials. In fact, on the contrary, economic systems have until now shown a tendency to evolve towards an increasing throughput of materials. Examining possibilities to reduce this throughput is therefore an important task in applying M-P chains.

CHAPTER 3

STRATEGIES AND POLICIES FOR M-P CHAINS

3.1. Introduction

This chapter presents various strategies and policies which may be employed for reducing the negative environmental impact of materials and products use. These strategies and policies will be used extensively in the chapters on modelling and applications in Parts II and III of this study.

The growing awareness of the negative environmental impacts of the use of materials and products has led to many environmental research efforts. A central theme in the present research is how to reduce these impacts on the environment. A distinction is made here between two main approaches to study policies and management aimed at reducing the environmental impacts of production and consumption. One is based on economic welfare theory, and the other is a more pragmatic, multidimensional approach. Section 3.2 deals with these approaches and the views on material and product flows. Section 3.3 presents the concept of chain management in the light of M-P chains. Section 3.4 then discusses the various strategies, such as recycling and substitution, which are available to reduce the environmental impacts of M-P chains. Section 3.5 examines environmental policies which may be suitable for materials and products. These policies are subdivided into regulatory, economic and persuasive instruments. They will be examined later in the applications in Part III of this study. Section 3.6 gives an overview of the practice of material and product policies in several countries, as agreed by various international organizations. Section 3.7 provides conclusions.

3.2. Alternative Approaches to Environmental Policy Evaluation

Environmental problems, policies and their evaluation may be considered from various perspectives. The main focus of environmental scientists is on the physical, chemical and ecological aspects and indicators of environmental problems. The focus of the mainstream economic literature on environmental problems may be divided into two perspectives. One is an economic welfare-based approach which is based on neoclassical welfare and equilibrium theory, in which the unpriced effects of economic activities are the central theme. This welfare approach is a monodisciplinary approach to environmental policy, generally aiming at optimizing. The other perspective may be referred to as a multidimensional approach as it explicitly includes environmental, economic, physical aspects and indicators. It may be characterized as a satisfying approach. It is also a multidisciplinary approach because it requires some degree of linking concepts and insights from various disciplines, such as environmental science, biology, physics and economics.

Welfare approach to environmental policy

In the neoclassical theory environmental problems or damage are considered in terms of 'externalities'. Various definitions of externalities can be found in the literature. The following one is well known and comes from the most influential text in this field: "an externality is present when an individual's utility or production function is affected by the behaviour of another individual who does not take the effects of his behaviour on the other individual into account" (Baumol and Oates, 1988). An externality is a type of market failure, due to which market processes do not lead to a socially optimal, Pareto-efficient allocation of scarce resources.

One may distinguish between technological and pecuniary (i.e. financial) externalities. Pecuniary externalities only affect the prices of inputs or outputs in the economy, but not the production function via direct or physical impacts as a technological externality does. Another distinction is into public and private, or depletable or divisible externalities. A depletable externality is 'consumed' by an individual so that it is not available any more for another individual. A public externality is non-depletable, e.g. the breathing of polluted air does not restrict the availability of this air to others. The same holds for a nice view in the mountains which does not affect the views of other persons (so long as the viewpoint is not too crowded). It has also been stressed that an externality is an unintentional or accidental effect of a legitimate action, which excludes deliberate or illegitimate actions, such as stealing, from being externalities. In neoclassical economics it is generally accepted that externalities occur due to inadequate property rights (Coase, 1960; Baumol and Oates, 1988; Verhoef, 1996).

By definition, externalities occur outside the market and therefore no market price is attached to them. Policies are needed to adjust for these external effects of production and consumption. Policies may correct for costs that are not included in the initial market price. The welfare optimizing theory studies policies designed to correct for an externality. In a situation with unpriced damage, a Pigouvian tax or levy may be imposed. This tax or levy is equal to the marginal damage costs in the social optimum. With this Pigouvian tax or levy the neoclassical welfare optimal (Pareto optimal) solution can be reached. The measurement or estimation of the optimal externality costs or marginal damage costs is difficult. There are many methods available for the estimation of external costs which often render different results (for an overview of methods, see Freeman, 1993; Hanley and Spash, 1993; Hoevenagel, 1994). The issue of such monetary valuation of externalities is not dealt with in this study. Externalities and the welfare approach are discussed in more detail for M-P chains in Chapter 5, namely in Section 5.3 on pollution economics, and in the general equilibrium model in Chapter 7.

A multidimensional approach to environmental policy

Many critics of the neoclassical externality theory argue that the optimization of environmental externalities is only possible in theory (e.g. Georgescu-Roegen, 1976;

Costanza *et al.*, 1990[1]; Daly, 1991 and 1996; Pezzey, 1992; Faber and Proops, 1992). Several reasons are offered in this respect. In practice, the optimal marginal damage costs are difficult to determine because of valuation problems. Besides the practical problems related to the estimation of external costs, there are also more fundamental problems. The optimal external costs are the external costs in the social optimum, while the estimate of the actual external costs will only give the non-optimal external costs. The inclusion of the external costs will give rise to other optimal costs. Furthermore, in practice the choice of a policy or instrument to optimally correct for an environmental externality does not only depend on economic efficiency, but also on various other evaluation criteria, such as environmental effectiveness, equity and environmental sustainability (e.g. biodiversity) and political feasibility (see Section 3.5.2). Especially the notion of sustainable development or environmental sustainability is not necessarily consistent with an optimal welfare approach (Van den Bergh, 1996).

Other market failures, such as imperfect markets, imperfect information and non-rational or strategic behaviour of agents, are often not considered in externality theory, where all individuals are assumed to be perfectly informed and optimize their behaviour (exceptions can be found in game theory and industrial economics). Other non-market failures, such as government, institutional (e.g. the lack of property rights) or transaction failures or the lack of altruism or anticipation, are not considered (Opschoor and Van der Straaten, 1993; Opschoor, 1996). All these arguments provide a convincing motivation for a distinct, more pragmatic, multidimensional approach to study environmental policies.

It has been stressed that a multidimensional approach to environmental problems should be based on several types of impact. The multidimensional approach may be more useful for practical reasons. It may be aimed at reducing the physical flows (throughput, Daly, 1996) or the environmental impact of the physical flows under a set of economic constraints. This reduction can be achieved via: (1) reducing the use of materials; (2) closing the chain of materials and products by recycling materials and the reuse of products; and, (3) changing to other types of materials and products. Note that in the latter case a shift in environmental problem(s) may occur, e.g. to problems related to other materials or mediums (see further, Section 3.4). This multidimensional approach may also be used for an optimization of costs under a set of physical or ecological constraints. It should be emphasized that in this approach economic aspects are included, so that it is different from environmental science or impact analysis as traditionally understood.

As in the theoretical economic approach, the reduction of the environmental burden may be realized by imposing environmental policies. The evaluation of these environmental policies is not only based on economic efficiency, but may also be based on other physical, ecological and economic effects, dimensions and criteria. For the ranking and evaluation of the various policies, multi-criteria analysis is one of the methods that may be used (Janssen, 1992).

---

[1] In Costanza *et al.* (1990) a more pluralistic economic approach to environmental problems is called 'ecological economics'.

The multidimensional approach needs to be based on the basic physical laws of thermodynamics and material balances, as described in Chapter 2. The basic concept used by economists is that there is a circular flow between producers and consumers. Goods and services flow from producers to consumers, and production factors from consumers to producers. From an orthodox economic viewpoint the material and product flows and their environmental impacts (externalities) are measured in monetary values (or utility), instead of being considered in physical terms or environmental impact dimensions. Although many economists are aware of the importance of physical flows they do not take them into account explicitly (Daly, 1996, p. 34). The welfare approach does not in fact even consider the environment explicitly, let alone real flows of materials and energy from and to it. Instead, it focuses on external effects as if directly occurring between economic agents. In the multidimensional approach policies may aim at reducing the physical or environmental problems as such, and not by translating these problems into (marginal) damage costs.

In summary, the main difference between the welfare (-economic, neoclassical) and the multidimensional approach is that the former approach optimizes the sum of all benefits and external costs, while the latter evaluates the policy instruments according to various criteria. Table 3.1 shows some differences between the two approaches.

Table 3.1. Main differences between the two approaches to environmental policy.

	Welfare approach	Multidimensional approach
Goal	Internalizing environmental damage, or optimizing the sum of private net benefits and external costs	Physical reduction of environmental damage
Measurement units and evaluation criteria	Monetary terms (external costs)	Various physical, environmental or economic terms

3.3. From Chain Analysis to Chain Management

Many studies dealing with material or product flows concentrate on one aspect of these flows, for example the reduction of the emissions of a specific material or substance. By looking at only one aspect, other (environmental) effects may be overlooked. Therefore, chain management has been introduced as an overall policy strategy that explicitly takes into account the (sequential) linkages between the various activities in such a chain, in terms of economic and physical mechanisms (Opschoor, 1994). Chain management can be defined as 'the management of material flows that result from networks of social and economic activities' (VROM, 1993b) or in terms of M-P chains as "the manipulation of M-P chains so as to optimize the environmental impact of these chains, or so as to achieve a certain accepted environmental impact at least social costs" (Opschoor, 1994). Chain management refers to the sequence of economic activities arising from extraction, production, consumption, recycling and waste-treatment processes. Environmental problems caused by the use, the flows and the accumulation of materials and products in all stages between extraction and waste treatment are considered simultaneously. The added value of chain management in comparison with traditional approaches, e.g. end-of-pipe technologies, is an increase in options to reduce environmental damage, and the coordinated tackling of problems at several points in the chain. The total environmental damage caused by the entire chain is the central issue, not the damage caused by just one specific stage of the chain. This is important because by only looking at one part of the chain, problems or effects that shift to other parts of the chain may be overlooked. The focus of chain management is not on one type of agent, such as consumers, but on various types which are often considered separately in traditional environmental management. In the Netherlands, the adoption of a chain management approach in all environmental policy making and by all economic agents is called 'integral chain management' (VROM, 1993b). Section 3.4 describes a variety of strategies which may be used to reduce environmental problems related to material and product flows.

M-P chain analysis may be very helpful for chain management because the effects of possible solutions may be analysed. M-P chain analysis considers the possible effects of changes in production or consumption on other parts of the chain. An M-P chain typically includes several producers and consumers, making chain management and analysis not limited to reducing the environmental damage caused by simply one agent.

M-P chain analysis examines the environmental, physical and economic aspects of product and material flows. The inclusion of economic aspects, such as prices, demand and markets, may complicate the policy analysis of the environmental problems, but these aspects do have a significant impact on policies and policy analysis. In M-P chains choices are made by various agents, such as producers, consumers and recycling firms. Such choices concern, for example, the use of materials in production and the choice by consumers for specific final goods to meet the demand for a service. These choices are often based on economic (price), social (fashion) and technical (durability) arguments, although environmental aspects (i.e. a 'green' image) may also play an important role. The purpose of M-P chain analysis is to investigate the size of physical flows, how these flows may alter in response to (endogenous) changes in the economic process, and how interventions can change the physical, economic and environmental aspects of these flows.

As discussed in Section 3.2, environmental problems can be considered using a welfare approach or a multidimensional approach, thus allowing for a comparative analysis. With an M-P chain analysis both approaches can be examined. In Chapters 5 and 6 they are discussed on the basis of various types of models. In Chapter 7 the welfare approach is applied in a general equilibrium model, while in Chapters 6, 8, 9 and 10 applications of the multidimensional approach are shown in a variety of models.

3.4. Strategies to Reduce Environmental Problems in M-P Chains

3.4.1. THE STRATEGIES

The use of materials and products may cause environmental problems. By changing the physical flows in M-P chains these problems may be reduced. Here, the (technical) strategies to achieve this reduction are listed and discussed in order to have a clear view on them in later chapters of this study. The following list of strategies may be used.

1 a Substitution between materials.
  b Substitution of a material by capital or labour.
  c Substitution between products.
2 a Recycling of materials.
  b Reuse of products.
3 a Technological change.
  b Change in product design.
4 Changing the pattern of consumption.

These various strategies are not mutually independent: for example, technological change may facilitate substitution, recycling or changes in product design, and recycling may encourage substitution between new and recycled materials. Each strategy will be discussed separately in subsequent sections.

3.4.2. SUBSTITUTION

Substitution can be defined as the replacement of one material by another material, or replacing one product by another product, without changing the function or use of the material or product. One may distinguish between different types or levels of substitution from the perspective of material and product flows. The three types of substitution will now be described in turn.

(a) *Direct substitution between materials in production.*

Direct substitution between materials means that in a production process one material input is replaced by another material input. This type of substitution may also be called inter-material substitution. Substitution of materials by less environmentally damaging materials does not make the M-P chain more cyclic but reduces the problems caused by it. This change in the use of materials is not necessarily driven by environmental policy but can also be the result of economic (e.g. scarcity) or technological factors (e.g. the application of techniques developed for other purposes). In automobiles the percentage of aluminium and plastics is increasing at the cost of iron and steel. This may be caused by the technical properties of these substitutes: for example, plastics are easier to mould than metals. For accounts of changes in the use of particular materials, see Eggert (1986 and 1990), Gjostein (1986), Tilton (1990), and Ginley (1994). Policies may result in forced substitution: for example, the ban on CFCs is resulting in the substitution of CFCs by other materials.

Materials substitution is important to reduce the environmental impacts of M-P chains in so far as using certain materials in a product or in a production process can reduce the use of other depletable or toxic materials. This type of substitution can also take place between new materials and recycled materials. In this case, the availability and the quality of recycled material are crucial factors.

Secondary effects, such as the shifting of environmental problems from one material to another, are an important factor to consider with materials substitution. Secondary effects may be seen as effects occurring as a result of a change aimed initially at having another (main) effect.

(b) *Indirect substitution between materials and non-materials in production.* The substitution between materials and non-materials, such as capital or labour, can be achieved by using more labour or capital, and less materials. This type of substitution may be called 'indirect' because it is substitution between different input categories. An example of indirect substitution between labour and materials is increasing labour input to check the production line in order to reduce the amount of waste generated, which will result in a more efficient use of materials. An example of using more capital and less materials is the use of computers (capital) to determine how to cut down on materials so as to obtain less waste material. This type of substitution is related to new process and product design. A secondary effect that may occur is that a new machine is needed for the new process, which may result in a quicker disposal of the old machine.

It is argued that substitution between materials and non-materials is limited. The limitation is illustrated by Daly (1996) using the example of baking a cake; baking a cake is impossible by unlimited substitution of flour and wheat by bigger ovens.

(c) *Substitution between products in consumption.* A third type of substitution is between products. Consumers or producers may replace one product by another that gives them the same service. The reason for such substitution is that the substitute is cheaper, more fashionable or 'greener'. Substitution between products may be affected by consumers, producers and policy makers (Boons, 1995). The replacement of one product by another one may have secondary effects for the environment: for example, the substitution of a mechanical product by an electrical one, e.g. coffee-machines. The secondary effect of this substitution between products is that more electricity is needed and possibly more materials to produce the machine.

An example of substitution between products is the change from tin-plate beer cans to aluminium cans (Nappi, 1986; Roberts, 1992). Because this change is closely related to technological change it will be further elaborated in Section 3.4.4.

3.4.3. RECYCLING OF MATERIALS AND REUSE OF PRODUCTS

Recycling is a topic that is discussed broadly by academia and governments. The various aspects related to recycling are discussed in this section. The reuse, repair and remanufacturing of products is discussed separately, because these may be seen as special types of recycling (Ayres and Ayres, 1996). The section ends with an economic view on recycling and reuse.

Recycling of materials
The recycling of materials can be used to reduce the extraction of materials and the generation of waste. Thus, in principle it has a favourable impact on the quality of the environment. Recycling is often subdivided in different types. *Primary recycling* is the recycling of materials within a production process (VROM, 1994). *Secondary recycling* is the recycling of materials after the product is used. Materials used for primary recycling are called 'new scrap' and for secondary recycling 'old scrap'. *Tertiary recycling* is the incineration of waste, which results in a recovery of energy. Some argue that incineration is not recycling, but just burning of waste. Another distinction is direct and indirect recycling. *Direct recycling* is the use of recycled materials for the same purpose as virgin materials. The possibilities for direct recycling depend on the quality and homogeneity of the materials. Most direct recycling takes place within a production process: for example, when broken glass bottles are used again to make new bottles. This is a sub-class of primary recycling. *Indirect recycling* is the use of recycled materials for a lower quality purpose (Porteous, 1996). An example is the use of recycled plastic for the production of traffic posts.

The possibilities of recycling are related to the production process and the materials used. The disassembly of products is more difficult when they consist of many different parts (and materials) or of very small parts of certain materials. An example is the use of plastics in household devices. Various types of plastic are used in one device which makes it difficult to collect and recycle those plastics. Therefore, it is better to reduce the number of different plastics in a single product to be able to recycle more. For example, in an average automobile there are between 45 and 115 different types of plastic, which are difficult if not impossible or too expensive to separate (Nijkerk, 1994). This implies that the use of less materials per product (dematerialization or miniaturization, see Section 3.4.4) is not in general preferable to the use of more of one type of material. The design of products will affect the possibilities of de-assembling for recycling. Therefore, at the product design stage the recyclability of a product should be taken into account.

Sometimes the recycling of materials can be technically impossible, because by corrosion and leaching the materials become too dispersed. An example is the lead in leaded gasoline which disperses into the air, so that it is impossible to collect the lead. In general it is considered that 100% recycling is not possible due to the second law of thermodynamics, or more accurately due to the 'fourth law of thermodynamics' (Georgescu-Roegen, 1971) (see Section 2.4. in Chapter 2).

Because recycling involves energy and materials, it is not necessarily an attractive strategy from an environmental perspective. Some life-cycle assessments for packaging have suggested that in some cases incineration is preferable to recycling (Huppes, 1993). However, most studies show that recycling requires less materials and energy than the extraction of virgin materials (see, for example Potier, 1977).

The paper industry is a good example of recycling. A part of the inputs of this industry is recycled paper. A high percentage of the wastepaper is recycled. For example, in the EU countries it ranged from 33 to 51 percent in 1990 (Bertolini, 1994). Recycled paper is extensively used by the paper industry because it is technically substitutable for new paper, and the production costs of both types of paper inputs are comparable. Waste paper recycling has been studied by Turner *et al.* (1977), Turner and Pearce (1994), and Van Beukering and Duraiappah (1996). The recycling and disposal of different forms of packaging waste has been evaluated in Brisson (1993). In the latter case, the policies of different countries were examined after which an analysis was performed of different charges on beverage containers for various countries. Another industry where recycling is an important issue is the automobile industry (Lévêque and Nadaï, 1995). The organizational and chain management issues of the recycling of cars have been studied by Den Hond (1996).

Quality of recycled materials

In general, recycled materials have a lower quality than new materials, but there are some exceptions: for example, recycled aluminium has the same quality as new aluminium. New and recycled aluminium cannot be distinguished and therefore no separate markets for new and recycled aluminium exist (Bekkers and Mulder, 1990). It is interesting that the production of secondary aluminium out of scrap only needs 5% of the energy required for the production of primary aluminium (Bekkers and Mulder, 1990). The reduction in quality of recycled materials is generally caused by contaminants in the materials (e.g. paint on metals or nails in wood), or to physical degradation (e.g. corrosion). Due to the lower quality, recycled materials are mostly used for other purposes which do not require high quality material. For example, the plastic in plastic cups is reused to make traffic posts. The continuous diminishing quality of materials via repeated recycling is often referred to as a 'cascade' of continually reused materials.

Resource cascading can be defined as the sequential exploitation of the full potential of a resource during its use (Van Elburg *et al.*, 1992; Sirkin and Ten Houten, 1993). An example of a cascade is the recycling of plastics. The following five stages chart the decreasing quality of the recycled plastics (Schneider, 1993): (1) bottle reuse; (2) mono-plastic recycling; (3) mixed-plastic recycling; (4) incineration with energy recovery; and, (5) landfill. The idea of cascading is relevant for many materials, for instance aluminium (Van Elburg *et al.*, 1992), wood (Sirkin and Ten Houten, 1993), pinewood (Fraanje, 1997a), and hemp and reef (Fraanje, 1997b). Tromp (1995b) gives a strategy for how the idea of the resource cascade may be used. The lower quality of recycled materials is just one of the reasons for preferring to use new materials. Other reasons may be that the law requires a certain percentage of purity which is difficult (technically or economically) to obtain with recycling, or that consumers or producers perceive secondary materials as inferior.

A special issue in recycling is 'closed-loop' and 'open-loop' recycling. Closed-loop recycling means that the recovered materials are reprocessed into the same product: for instance, the aluminium of aluminium cans is recycled and used for such cans again (McClain, 1995). It is less common than open-loop recycling, where the recovered materials are used in another (type of) product. Closed-loop recycling can give specific problems. An example in agriculture concerns vegetables which may contain a small amount of toxic elements that end up being consumed by households. A part of those vegetables is collected from the household waste to become compost which is then used to fertilize the ground. This organic household waste may contain these toxic elements, such as cadmium, which are dispersed in the ground with the compost. These elements are not biodegradable so that they accumulate in the soil, and may ultimately damage crops and endanger human health (Moolenaar *et al.*, 1997).

Reuse, repair and remanufacturing of products

The reuse of products is closely linked to the recycling of materials, but it differs in the sense that for materials recycling the products need to be disassembled or converted by physical or chemical processes into reusable materials. By contrast, the reuse of products may simply require that the products are just cleaned: for instance, glass milk bottles. Thus, new and reused products have not changed either physically or chemically.

In general the quality and the economic value of reused products is lower than that of new products: for instance, second-hand clothes or furniture have a lower quality and value than new clothes and furniture. The cascade idea described above is also applicable to the reuse of products. Sometimes the quality may decrease but that is not relevant or visible to the consumer: for example, for the purchase of glass milk bottles it is not relevant whether the bottles are new or reused. The producers take care that the quality of the bottles is sufficient. Consumers and producers need to have confidence in the quality of reused products. Technically the products have to be fit to be used again. Due to the decrease in the quality of the products after prolonged use they cannot be reused infinitely.

Furthermore, innovation of products may mean that the original version becomes economically obsolete. In general, the economic lifetime of a product is shorter than its physical lifetime. An example of extremely fast innovation and modernization is the computer industry in which the development of computers goes so fast that after only a few years current models of computers become economically obsolete.

Repair of products is declining, largely because products have become more complex which makes repair difficult, expensive and time-consuming (Ayres and Ayres, 1996). Many producers prefer to sell new products instead of selling components useful for repairing products.

However, the remanufacturing of products may become more important, because of the difficulties associated with waste disposal. For some products a take-back requirement is put in practice, which obliges producers to take back their product after it is used and disposed of by consumers. Remanufacturing is very labour-intensive, but it may pay off: for example, in the case of products like computers and refrigerators. Especially in countries with low labour costs the remanufacturing of products is economically attractive, while in countries with high labour costs, a lower VAT on recycling or repair activities may need to be used as an incentive for more recycling and repair.

An economic view on recycling and reuse

Apart from technological restrictions on recycling and reuse, economic constraints are essential too. The first requirement for recycling and reuse to take place is that a market exists for recycled materials and reused products. Due to the fact that recycled materials and reused products usually have a lower quality (and value) than virgin materials and new materials, recycled materials and reused products are only bought when they are less expensive than new materials and products. For the firm to start recycling or reuse, the costs of collection, cleaning, recycling of materials or reuse of product have to be less than the revenue from the recycled materials or the reused products. Thus, although recycling and reuse are environmentally attractive and technically possible, economic factors, such as the costs of recycling or the lack of demand, may make these options unattractive.

When considering the economic efficiency of recycling from a welfare economic approach, as described in Section 3.2, it does not consider recycling as either good or bad, but instead it examines the market failures or the non-efficiency of the market for recycled and new materials (Page, 1977). One of the market failures is that the external costs for society are not included in the costs of extracting new materials and recycling materials. The effect of including the external costs of disposal is analysed for the optimal level of recycling.

The costs of recycling for individual consumers and for society are different. When a consumer has a product that is no longer of any use to him, he decides either to recycle it by bringing it to the recycling firm, or to dispose of it the garbage bin. For the consumer, the costs of recycling are higher, e.g. the transport costs of bringing the product to a recycling centre, than the costs of disposal. Especially when the costs of disposal are a fixed rate (i.e. independent of the amount of waste), the marginal costs of adding more waste are zero. Figure 3.1 illustrates that the optimal percentage of recycling is different for society (the marginal societal costs of disposal, SMCd) and for a consumer (the private marginal costs of disposal, PMCd). The market level of recycling (Qp), i.e.the level where the marginal costs of recycling (PMCr) equal the marginal private costs of disposal (PMCd), is lower than the optimal societal level of recycling (Qs).

A government may impose policies, such as subsidies on waste paper collection, to stimulate recycling and reuse. With such policies the market level of recycling may be equal to the optimal societal level of recycling (see Figure 3.1). An issue in recycling is the trade in recyclable materials between developed and developing countries: for example, the export of discarded refrigerators and car tyres to Africa and of waste paper to India (Van Beukering, 1997). More on the social and economic dimensions of recycling and reuse can be found in Pearce and Walter (1977).

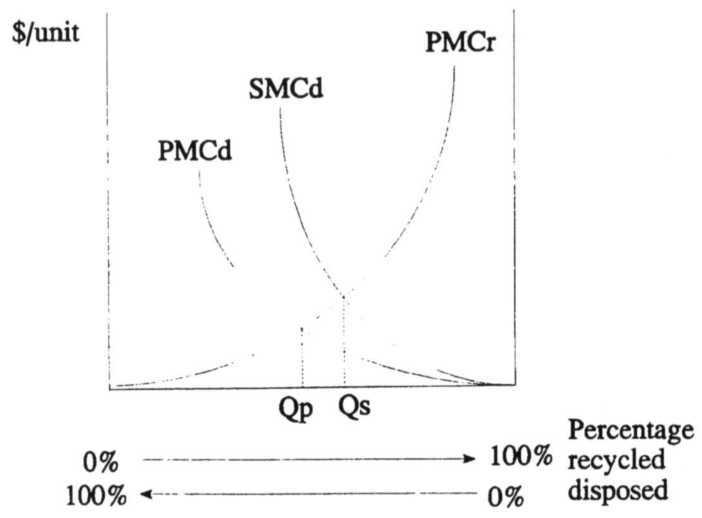

Figure 3.1. The optimal level of recycling (adapted from Tietenberg, 1996, p. 181).

3.4.4. TECHNOLOGICAL CHANGE

This section discusses several types of technological change, based on a classification of technologies by Kemp (1995). First, pollution control technologies that prevent the direct release of environmentally hazardous emissions into the air, water or soil, i.e. end-of-pipe technologies, are described briefly. Second, process-integrated changes in production technology, input material changes and good housekeeping are examined. These reduce the amount of pollutants and waste material that is generated during production, and diminish the usage of environmentally harmful materials. Third, technological change directed at products is discussed. Technological change, such as product redesign, may result in new products that substitute for existing products, and that have a lower environmental burden: for example, zero-emission motor vehicles and low-solvent paints.

In M-P chains technological change can (i) facilitate substitution and recycling in order to produce and consume in a more environmentally-friendly way; (ii) lead to changes in product design in terms of less use of materials; and, (iii) lead to the use of other products or materials to fulfil the same demand.

Another division of technological changes are radical or gradual processes (Kemp, 1995). Radical or discontinuous changes may result from sudden changes: for example, in the environment or in policies. Gradual changes are continuous processes which result from small modifications and adaptations of technological systems, e.g. modifications of existing products or the use of end-of-the-pipe technologies. These changes may be important from an economic perspective, but they do not change the production process fundamentally.

Changes in end-of-pipe technologies

The first reaction to environmental problems was to look at the emissions and waste of firms at the end of the production stage, where pollution problems were most visible. This is called end-of-pipe treatment. This may involve, for instance, introducing filters on chimneys in order to emit less emissions or less pollution. Many different technologies have been invented and used for this end-of-pipe treatment. However, they have not resulted in reducing or eliminating waste, only in merely redirecting the waste. 'Sequestration', which is the disposal of waste on a controlled disposal site, may be seen as a type of end-of-pipe treatment (Ayres and Ayres, 1996).

Change in process design

As explained above, end-of-pipe technologies do not reduce the waste and emissions of a production process, but merely redirect them. There is thus a need for changes in the process design or process technologies. Process change starts with the invention of a material, a production process or a product, after which innovation of technologies and diffusion of those technologies may take place. Factors that may drive process change are competition (economic reason), legislation (governmental policies), and the environment.

Often technological changes are necessary to fulfil the governmental requirements: for example, regarding environmental standards. Governmental policies may require that the firms have to reduce their emissions: for example, firms may need to install a new filter on their pipes (end-of-the-pipe technology) or to implement a new production process (process-integrated technology).

The effects of technological change on the use of materials may have important economic and environmental effects: for example, the use of new materials may be economically and environmentally beneficial. An illustration of this is the use of materials for beer cans. Until 1964 there were only glass bottles and tin-plate cans in the United States, but in 1964 aluminium cans were introduced in the market, after which the steel producers made a tin-free steel can to compete with aluminium cans and glass bottles. By 1982 almost no steel cans were produced. Glass bottles lost a part of their share of the market to aluminium cans but still have around 30 percent of the beer container market. Besides changes in the type of materials used also the weight of the cans changed significantly. The weight of a standard steel can (12 oz) was 44 grams in 1960, 34 grams in 1970 and 27 grams in 1985. The weight of an aluminium can decreased from 25 grams in 1964, to 20 grams in 1970 and to 17 grams in 1985 (Roberts, 1992). Figure 3.2 presents the market shares in the beer container market of the various types of containers over time.

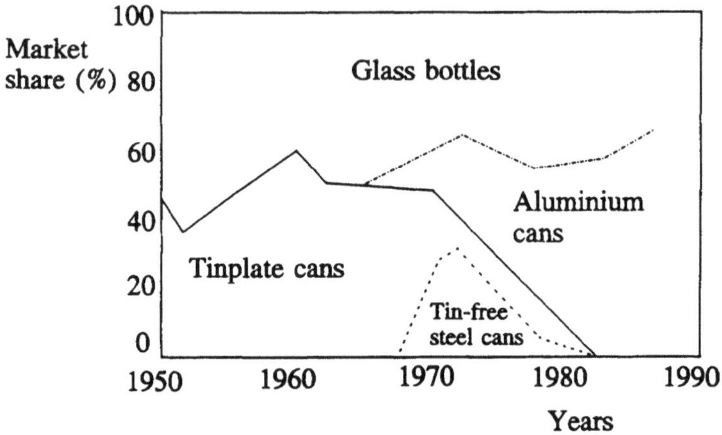

Figure 3.2. The shares of various types of containers in the beer container market per type over time (Roberts, 1992).

Changes in product design

In this section the changes in the product design are described. The changes in product design are on the producers' side of the market but they are closely related to the demand for products (see Section 3.4.5).

Products can be made less environmentally damaging by changing the product design. Firms can look for ways to improve the product design of existing products or can make a totally new product. Developments in technology may change product design options (Moll, 1993). Changes in product design that are beneficial for the environment are the following. A first type of change is a reduction in the use of materials in a product, also called dematerialization (see Chapter 2, and Herman *et al.*, 1989). The use of less materials or other materials may have negative effects on the environment too: for example, the use of combined materials or smaller quantities of materials may hinder recycling, because it is technically more difficult or more expensive to separate the materials. Tromp (1995a) uses the term 'miniaturization' for using smaller quantities of specific materials in products: for example, the use of heavy metals in computers. A second type of change is the use of less environmentally damaging materials. The use of other materials in the design of a product may be beneficial for the environment when less environmentally damaging materials are used or recyclable instead of non-recyclable materials. An example is the use of wood instead of metals or the use of paper instead of plastic. Third, a longer life cycle of the product generates less waste (Conn, 1977). An example of a prolonged life cycle is the introduction of rechargeable batteries which extends the lifetime of batteries. It is important, however, to take into account the delayed effect, because in the long run these batteries are also disposed of, and this waste needs to be collected and recycled or otherwise treated.

3.4.5. CHANGING THE PATTERNS OF CONSUMPTION

The strategies described above are mostly related to the producers. However, consumers decide which products to purchase, how they treat a product in use (the maintenance), whether to discard or repair the product, and whether to purchase new or reused (second-hand) products. Changing the pattern of consumption is a crucial factor in reducing or altering the material and product flows. The consumption pattern may be altered by changing prices, the availability of substitutes, or preferences.

The demand for a product may change when substitutes are available that give the same service. The availability of substitutes depends on technological possibilities, the design of products, but also on economic aspects such as the costs of extraction, recycling, reuse and dumping. The price differences between substitutable products partly determines which product consumers will buy. The price of products may be influenced by public policies, for example a tax on leaded gasoline in many OECD countries (OECD, 1994).

In addition, consumer preferences, focusing on fashion and durability of the product, are part of the decision to purchase a specific product. Consumer and producer preferences may change over time and can be influenced by information or education about certain products. An example is the choice between bleached and unbleached paper, which has been promoted by information from governmental and non-governmental organizations. Another example is the stimulation or promotion of the use of pinewood window frames by the city of Amsterdam which has resulted in an increase from 1% to 52% in the use of pinewood window frames in two years. The use of hardwood, aluminium and pvc window frames has decreased sharply (Fraanje *et al.*, 1992). The change in demand after the introduction of a new product can be illustrated by 'high-efficiency boilers' (Brezet, 1994).

The strategies described in Section 3.4 are not all compatible with each other. For instance, the effect of miniaturization of materials which are used in products is twofold (Tromp, 1995a). On one side, it is beneficial for the environment because less materials are used, but, on the other side, the smaller parts are more difficult to repair, disassemble and recycle. Sometimes miniaturization or dematerialization results in the use of other, more advanced, but possibly less recyclable materials.

Changes in products may also result in the use of more materials, because the demand for those (new) products may increase, and also more capital may be required for the production process. Another important effect that may occur due to a change in products is that the older products are disposed of more quickly. This disposal results in more waste. A good example is the rapid changes in computers: a computer that was the fastest and most modern 3, or even 2 years ago, is now outdated.

3.5. Environmental Policies for Chain Management

3.5.1. ENVIRONMENTAL POLICY INSTRUMENTS

To reduce environmental externalities caused by the production and consumption of products environmental or optimal policies may be imposed. Various policies exist which can be classified as (i) regulatory or command-and-control, (ii) economic (or market-based), and (iii) persuasive (or social) instruments. Below, these types of instruments will be discussed separately.

(i) Regulatory or 'command and control' instruments

Regulatory instruments include standards with which polluters have to comply. If the rules are not obeyed a penalty may be imposed. There are four categories of standards (Barde, 1995): (1) ambient quality standards: for example, the maximum level of sulphur dioxide in the air; (2) emissions or discharge standards: for example, the maximum output of SO_x emissions by an industry; (3) process standards: for example, the obligatory installation of filters in chimneys; and, (4) product standards: for example, catalysts in cars.

Regulatory instruments can be imposed on materials used, product features, production processes and waste treatment processes. Ambient quality standards, emission standards and process standards can be imposed on the emissions of a production process of the materials or goods. More specifically related to material and product flows are standards on recycling or recycled materials: for example, a fixed percentage of the materials in the production process have to be recycled. A process standard might be that the materials used for a product are standardized to facilitate recycling after the product is disposed of. Product standards can be imposed to reuse a certain percentage or number of products or to design products in such a way that recycling the materials is easier. Regulatory instruments may be imposed on waste treatment processes, in terms of the type and quantity of waste or emissions: for example, protected storage instead of dispersion into the air. Thus, for M-P chains regulatory instruments may be imposed in many different stages of the chain: from new materials use to waste treatment.

The assessment of the standards can be based on the best available technology, but these assessments do not give an incentive to improve the technologies for abatement, product design or recycling. Instead, stricter standards in combination with a time horizon can be used as an incentive for R&D because they stimulate producers to innovate and to adopt the new, cleaner technology. These stricter standards might be imposed with a time-lag to allow producers to develop and adopt alternative technologies.

(ii) Economic instruments

Economic or financial instruments are imposed to directly change the costs and benefits associated with decisions and the behaviour of producers and consumers (Baumol and Oates, 1988; OECD, 1989; OECD, 1994). Economic or market-based instruments may be imposed at various stages of the M-P chain. Here, the types of instruments and the possibilities of imposing them in M-P chains are examined.

Economic instruments leave economic agents considerable freedom to respond to changes in prices. This is in contrast with regulatory instruments that force agents to change particular decisions. The behaviour of agents may be changed by making the environmentally less-damaging alternatives more attractive. Economic instruments may also be used for changing the base of taxation, from a tax based on labour and capital towards a tax based on material or energy inputs. This fiscal reform aims to use less materials and energy, and to recycle more. Such tax reform is sometimes claimed to have a so-called 'double dividend' because less materials are used (better for the environment) and more employment is generated (Bovenberg and de Mooij, 1994; O'Riordan, 1997; Bohm, 1997).

The basis of the tax level can often not be measured directly and therefore approximations are used: for example, the carbon content of fuels as a proxy for the CO_2 tax. In a neoclassical framework economic instruments are used to internalize the external costs related to production and consumption. These are called Pigouvian taxes (see, Baumol and Oates, 1988). Potential environmental policy instruments are taxes, charges, levies, deposit-refund systems, subsidies, marketable permits, property rights and tax differentiation.

The use of economic instruments has so far been rather limited compared with direct regulation for several reasons (Opschoor and Turner, 1994; Opschoor, 1995). First, with regulations the environmental goal is fixed, provided that there is enough enforcement, but with economic instruments it is more complicated to set the rate of the tax to obtain a certain environmental goal. Second, policy makers look at the short-term effects of policies and therefore ignore the long-term and indirect effects of regulations and other instruments. Third, taxes may be costly and technically and administratively difficult to impose. The use of subsidies is not very popular because these affect the governmental budget negatively.

Taxes, charges and levies

A pollution tax, charge or levy can be defined as 'a payment for each unit of pollutant discharged into the environment or for each unit of environmental damage' (Barde, 1995). The advantages of taxes are threefold: the total abatement costs are minimized; an incentive to reduce pollution is given; and a revenue for the government is provided. One difficulty of taxation is choosing that rate of the tax which achieves the environmental objective. Another negative aspect may be the distributive impact of a tax: for example, the burden of a tax on energy will be carried more by energy intensive industries. However, the distributive impact of a tax is not necessarily negative: for instance, when a tax is imposed on the use of gasoline.

Charges may have an incentive or a revenue-raising impact (OECD, 1989). The goal of a charge with an incentive impact is to change the behaviour of producers and consumers by raising the price of some products. Most charges have a revenue-raising impact that may be used for specific (environmental) uses (earmarked charges) or for general use. The revenues of regulatory charges are redistributed to the tax payers. The difference between a tax and a charge is that a charge is a payment for which a service is given in return (for example, the charge for municipal waste collection), while a tax is added to the government budget without giving a direct service in return.

Taxes and charges can be used in the context of chain management on the use of specific (raw) materials or products. In an M-P chain taxes can be imposed at the extraction, production, recycling or waste treatment stage or for a mixture of these stages. From a welfare approach a tax on, say, a virgin material may imply that the external costs caused by using this material are optimized. This optimization of externalities is also called 'internalization'. From an environmental viewpoint, a tax on a virgin material may cause a substitution from virgin to recycled materials.

Tradeable permits

In a system of tradeable permits, the government sets a standard for the 'optimal' amount of permits and issues permits so that the total quantity of pollution is restricted and the total costs of realizing that quantity is minimal. The idea was first published in Dales (1968). These permits are distributed or sold to polluters. In the theoretically ideal system the government does not have any costs and the polluting firms gain from these permits (Barde, 1995). The market for tradeable permits sets the price, which is automatically adjusted to all sorts of changes, including technical progress, entry of new firms, and inflation. One disadvantage of the system of tradeable permits is that is difficult to assess the level of the standard in order to achieve the environmental goal. Another difficult issue is the initial allocation of the permits to agents. Moreover, the monitoring of the transactions and the transaction costs may reduce the trade of permits (Stavins, 1995). Hence, transaction costs may reduce the cost-effectiveness of the tradeable permit system. The possibility of transferring the pollution to other regions is important, as also is the opposition against allowing the market to deal with environmental issues.

Tradeable permit systems have attracted a wide interest because they may offer advantages, such as a fixed environmental objective that does not change over time compared with other market-based instruments (Tietenberg, 1995; Barde, 1995). In some countries tradeable permits are already being applied, mainly for controlling air pollution (see Section 3.6).

In the context of M-P chains tradeable permits for the use or manufacture of specific materials or products could be a strategy to reduce the use of new materials by using less materials or by using recycled materials instead of new ones.

Subsidies

A subsidy is a payment to a producer or consumer for environmentally less-damaging acts. Theoretically, it is possible to achieve the same results with a subsidy and a tax (Bohm, 1981). The disadvantages of subsidies are that: (1) the costs of subsidies have to be carried by the government (i.e. general taxes); and, (2) in the case of perfect competition, subsidies do not necessarily reduce the emission of a firm nor of the whole industry because the entry of new firms is stimulated (Baumol and Oates, 1988).

A subsidy contravenes the Polluter Pays Principle, according to which the polluter should be made to pay for pollution reduction up to pre-determined levels instead of receiving a reward for such efforts (OECD, 1972; Netherlands Scientific Council for Government Policy, 1992). In practice, however, tax reductions are still used and in fact such systems of positive incentives are again coming into favour with policy makers (Werkgroep Vergroening Fiscaal Stelsel, 1996). An example is the tax reduction for investments that are beneficial for the environment.

Subsidies can be imposed at different stages of the M-P chain: for example, a subsidy on recycling or reuse and a subsidy on specific materials or products to make them more attractive to producers and consumers. A process subsidy can also be given to implement a new technology, which reduces the materials used or emissions. Seen from a M-P chain perspective, subsidies may be used to eliminate obstacles to achieving reduced or altered physical flows. For example, if there is no market for recycled materials, materials that may technically be recycled after consumption will not be recycled. When a market for recycled materials does not exist due to the availability of less expensive new materials, then a subsidy may be given to the recycled materials to stimulate their use: for instance, subsidies on used paper to encourage its collection.

Deposit-refund systems

A deposit-refund system is essentially a combination of a tax (deposit) and a subsidy (refund). A deposit is imposed which is later refunded as a subsidy when producers or consumers fulfil certain conditions. For example, a deposit is paid on a glass bottle and when the consumer returns the glass bottle to the shop he obtains a refund. This combination makes the system financially neutral, i.e. no government funds are used and no distributional effects occur. A deposit-refund system stimulates activities that otherwise would not have been undertaken (Bohm, 1981). A deposit-refund system has two effects: the reduction of materials or product use, and the increase of recycling and reuse. Therefore, it is an attractive strategy to apply to materials and products which are recyclable or reusable.

Deposit-refunds can be imposed on products and part(s) of products (e.g. packaging). These products may then be either reused (glass bottles), or recycled. A more recent theme in chain management is the use of deposit-refund systems for materials or substances (Huppes, 1993; Opschoor, 1994). These can be especially useful for materials which are environmentally damaging: for example, heavy metals, sulphur, nitrogen and phosphor. If a certain material is needed for a product, a deposit has to be paid, depending on the amount of material used. This deposit is returned when the product is offered for waste management or exported. A deposit-refund system may be extended to a generalized deposit-refund system (or a combination of a tax-subsidy system), where the agent who pays the deposit is other than the one who receives the refund. For example, the producer pays a deposit or tax for using new materials and the consumer receives a refund or subsidy when the product or material is returned in good condition, while the price of the product is higher due to the deposit. This may be interpreted as a generalized deposit-refund system.

(iii) Persuasive or social instruments

Persuasive or social instruments may be used to try to change the behaviour of consumers and producers by changing their preferences, opinions or values. Here, the persuasive or social instruments are interpreted broadly, including education, information, voluntary agreements (covenants), training, social pressure and negotiations.

Voluntary agreements
Voluntary agreements can be defined as "deals between government and industry, whereby an industry sector or a group of individual corporations agrees to reach certain environmental objectives within a defined time frame" (Barde, 1995). The advantages of voluntary agreements are that they are flexible and that industries themselves are taking the responsibility, although this responsibility is mostly taken because of the threat of governmental legislation or policies if the agreement fails to become effective. The negotiations between government and industries or amongst industries may combine the knowledge on technologies and costs of several firms. If the goal is not reached the government may impose other policies in order to reach it. In M-P chains voluntary agreements may exist, for example between producers who agree to take back their products after they are used. This may result in more recycling and reuse of discarded products.

Since the 1980s some voluntary agreements have been made, for example the French 'Eco-emballage' for packaging, and the Dutch 'packaging covenant' (Brisson, 1993). Organizations exist that are financed by a charge on packaging, which enables them to control the packaging chain, e.g. collection and recycling of packaging waste. Examples of such organizations are the German 'Duales System' (see Section 3.6).

Social regulation: information and education
Information and education given or supported by the government are mostly designed to change the behaviour of consumers and producers. A requirement is that the information provided is not controversial. An example is 'ecolabelling' that allows consumers and users to choose a product on the basis of information supplied about the environmental aspects of a product (Oosterhuis *et al.*, 1996). Products that fulfil a range of requirements are given a label. Thus, all information about the environmental or other aspects is confined to one label. This label summarizes for the consumers all the information about a product in order to allow the consumer to choose between that product and an (un)labelled substitute. In the Netherlands it has been discussed whether to label soya beans and soya-based products, to indicate whether or not the soya beans have been genetically manipulated.

For M-P chains information on products or the materials used in products may be used to allow consumers to choose between various products on the basis of environmental criteria. Some consumers will change their consumption (or preferences) when they are better informed about products. For example, a few years ago coffee filters were white but most consumers did not then know that chlorine was used to bleach these filters. Once this bleaching procedure became widely known unbleached brown filters were increasingly adopted. Another example is the introduction of separate bins for organic waste materials. Although consumers do not financially profit from collecting their organic waste separately, they do it voluntarily or under pressure from their social environment.

Which type of policy will be implemented depends on many factors, such as the number of polluters, the number of victims, the location of the polluters and victims (local, regional, fluvial, national, continental, global), the type of market involved, spatial features (diffuse or point-source pollution, and transboundary or global problems), vested interests, political processes, and international agreements and policy coordination.

3.5.2. CRITERIA FOR EVALUATION OF POLICY INSTRUMENTS

The choice of instruments for environmental policy depends on various criteria (Siebert, 1987; OECD, 1989; Barde, 1995). The three main categories of criteria below shows the complexity of such a choice.

1. *Effectiveness and feasibility criteria*
   a. Environmental effectiveness: Does the instrument help in reaching the environmental goal? Does it induce abatement? Does it take into account uncertainties about natural processes, economic data, exogenous developments and strategic behaviour of agents? Is it effective in the face of various changes and developments in the future?
   b. Technological feasibility: Can the technologies available or to be developed realize the objective?
   c. Sustainability criteria: Is the policy successful over a longer period of time and does it contribute positively to the welfare, environmental sustainability and well-being of future generations?
2. *Economic criteria*
   a. Economic efficiency: Is social welfare optimal or is the allocation of materials or products Pareto optimal? Are abatement or damage costs equal to the marginal costs?
   b. Cost-effectiveness: Are targets reached at least costs?
   c. Management and enforcement costs: What are the costs of implementation and control?
3. *Political criteria*
   a. Distributional and equity effects: Does the instrument generate acceptable distributional effects or can undesired equity effects be compensated by additional measures?
   b. Acceptability: Is the instrument acceptable for the industry and society?
   c. Simplicity: Is the instrument easy to implement and to enforce?
   d. Political acceptance: Is the instrument acceptable for politicians and their voters?

In the applications of M-P chain analysis in Part III effectiveness, feasibility, and economic criteria will play the most important role. In particular, the evaluation of applications based on the multidimensional approach (Chapters 6, 8 and 9) is in terms of a variety of environmental criteria (e.g. acidification, energy use) and costs for consumers, producers, regulatory agencies or an 'environmental manager'. Chapter 7 focuses on economic efficiency, i.e. the social welfare optimization, and Chapter 10 on welfare optimization of consumers and producers.

3.6. Policies Focusing on Materials and Products in Practice

Various policies have been imposed on materials and products. This section is meant to give an idea of the applications of policies in reality. It is not intended to give a comprehensive survey of all existing materials and product policies. The section is divided in three parts: the Netherlands, other countries and international organizations.

The Netherlands

The Dutch product policy focuses simultaneously on the main issues of the Dutch environmental policy. These issues are: (i) climate, i.e. the reduction or stabilization of CO_2 emissions; (ii) acidification; (iii) waste prevention, recycling and reuse; (iv) cleaning up of polluted soil; (v) reduction of energy use; and (vi) mobility (VROM, 1990). In the product policy the goal is to: 'reach a situation in which all agents - producers, consumers and retailers - continuously aim to reduce the environmental burden of products' (VROM; 1990, 1993b). The product policy aims at:

- the optimal use of (non-renewable) materials;
- the minimal use of energy in the entire product chain;
- innovation of products to increase their lifetime and to make repair easier; and,
- reduction of emissions and waste to an acceptable level during the entire product chain.

Since the Dutch environmental policy of 1989 consumers or households have been regarded as a relevant group of actors at which to direct environmental policy (VROM, 1989). Consumers determine the type and the quantity of products and therefore are the main focus of producers and possibly also of the policy makers. In M-P chains consumers are an important agent in the cessation or reduction of those flows.

The government intends to stimulate the education and information of consumers vis-à-vis the impact of their actions on the environment. Those actions are, for example, the purchase and use of goods, and waste treatment. The goals of the government are to collect many different kinds of waste material separately: used batteries, small chemical waste, glass, textiles, paper, metal cans, organic waste.

In the Dutch environmental policy plan of 1990 (NMP+) waste prevention, recycling and reuse was promoted by: looking at strategies for waste prevention and recycling in selected waste streams; a product policy; promoting recycling by quality control of secondary materials; using instruments like environmental care and permits; and, where possible, deposit-refund systems and regulatory levies (VROM, 1990).

The product policy of the Dutch government is based on integral chain management (see Section 3.2). This integral chain management is a strategy of the Dutch government which forms the basis of all its environmental policies, and which tries to optimize the environmental burden of the whole chain right through from extraction to waste treatment (VROM, 1993a). This strategy is a multidimensional approach to environmental policy (see Section 3.1).

In the Dutch product policy various product groups have been selected in order to determine their environmental impact during their whole life cycle. The selection of those product groups is based on products that generate environmental impacts in more than one compartment (air, water, soil) or during more than one stage (production, use, waste treatment). If and when data are available over the entire life cycle, the stages during which the most substantial environmental effects occur can then be identified. To improve the environmental performance of the product groups, all instruments, both regulatory and financial, may be used to diminish the environmental impact. Amongst the product groups selected are products of the chemical sector; leather products; paper; textiles; refrigerators; televisions; communication devices; washing machines; heating devices; lights; and, furniture.

The use of economic instruments in the Netherlands has not been widespread. Charges are used for municipal waste and sewage treatment, but these are mainly per household and not per kilogram of waste. Some municipalities have introduced a municipal waste tax per kilogram. Tax differentiation applies for the purchase of certain features of automobiles, and for leaded and unleaded gasoline. Deposit-refund systems are employed for glass bottles and some plastic bottles. Subsidies are given to industry for energy savings, reduction of heavy metals in water waste and reduction in the use of PCBs. The charges, tax differentiation and deposit-refund systems have been very successful, but the subsidies have only had a limited impact (OECD, 1994).

In the Dutch Packaging Covenant the packaging industry has voluntarily agreed on the conditions to reach a very detailed list of goals. The covenant aims at reaching these goals through standards, i.e. regulatory instruments, and not with economic instruments that are based on marginal benefits and costs. For a short, but comprehensive overview of this Covenant see Brisson (1993).

The focus of the Dutch environmental policy has been rather on direct regulations than on economic instruments. In the national policy plan of 1989 it was stated that the use of economic instruments only needs to be studied in cases where the regulatory instruments are not sufficient (VROM, 1989). However, only one year later, the new plan stimulated research on the use of some economic instruments: for example, a levy on primary resources, deposit-refund systems, materials-deposit-refund systems and levies on energy and minerals (VROM, 1990). In the 1993 second national environmental plan, the economic instruments proposed were mostly fiscal measures (VROM, 1993c). The only levy on materials and products that was proposed was a levy on energy for small-users which was imposed in 1996. It can be concluded therefore that economic instruments are still rarely used. Nevertheless, Opschoor (1995) expects that economic instruments will be used more, because regulatory and persuasive instruments are turning out not to be sufficiently effective, and because economic instruments may be more efficient.

Other countries

The following description of materials and product policies is divided into two main parts. The first part deals with economic instruments for product and materials policy and is partly based on an existing survey on the use of economic instruments (OECD, 1994). In line with this survey the categories of instruments that are distinguished are charges on emissions and on products, deposit-refund systems and tradeable permits. The second part deals with the command-and-control instruments that are used.

Economic instruments

Charges on emissions are either measured directly or indirectly at the source, or as a flat rate which means that every kilogram of emission is taxed at the same rate. Directly measured charges are, for example, the NO_x charge in Sweden. An example of an indirect charge is the charge on waste water treatment which is measured by the total amount of water used. A flat rate is used for municipal waste charges. Charges on emissions are levied on air pollution, aircraft noise, soil protection, waste, waste disposal, hazardous waste, and emissions to surface water.

Charges can be levied on materials, e.g. raw materials, or on products, e.g. batteries, on product characteristics, e.g. carbon or sulphur content. Also a tax differentiation can be included in this category of charges. Oosterhuis *et al.* (1996) systematically analyse various product policies in Europe. Table 3.2 shows some product and materials charges which are imposed in OECD countries.

Table 3.2. Charges on materials and products in OECD countries (adapted from OECD, 1994).

	Carbon	Sulphur in oil	Fertilizer	Pesticides	Batteries	Lubricant oil	CFCs	Pack-aging
Australia							x	
Austria			x					
Belgium								x
Canada								x
Denmark	x				x		x	x
Finland	x		x			x		
France						x		
Italy	x					x		
Netherlands	x							
Norway	x	x	x	x		x		x
Portugal					x			x
Sweden	x	x	x	x	x			x
USA			x				x	

Additional product charges are imposed on different products: on cars or car parts, such as: air conditioners in cars, in Austria, Canada, Denmark, Finland, Greece, Sweden, USA; diapers (Canada); plastic shopping bags (Italy); conventional light bulbs (Denmark); raw materials (Denmark). A tax differentiation is applied for the purchase of cars, the annual automobile tax or the sale of leaded as opposed to unleaded gasoline (most OECD countries).

Deposit-refund systems are either imposed by the government or industry. When imposed by industry such systems can be assumed to be economically profitable. There are deposit-refund systems for discarded car frames (Greece, Norway, Sweden); metal cans (Australia, Canada, Portugal, Sweden, USA); plastic beverage containers (Australia, Austria, Canada, Denmark, Finland, Germany, Iceland, Norway, Portugal, Sweden, USA); glass bottles (most OECD member countries). Other deposit-refund systems apply for fluorescent light bulbs (Austria); detergent packaging, drinks and dispersion paints and beverage containers (Germany); and, vehicle batteries (USA).

Tradeable permits systems are not used widely. Some examples of tradeable permits are those for acid rain control (Canada, USA) and ozone depletion chemicals (Canada, USA). Tradeable credits are given for low emission vehicles (USA), salt reduction (Australia), and oxygenated gasoline (USA) (Tietenberg, 1995).

Command-and-control policies

An example of regulatory materials and products policies is the 'Dual System' that exists for packaging waste in Germany. In that country a packaging ordinance was imposed in 1991. This states that packaging should be reduced to the minimum necessary for the product and should be refillable as far as that is technically possible and economically feasible, otherwise it should be reprocessed (Brisson, 1993). The 'Dual System' is a system set up for the collection and reprocessing of packaging waste to fulfil the requirements of the German packaging ordinance. The Dual System issues licences for which producers pay a fee and can then place a 'green dot' on their products. These fees are used to finance the collection, recycling and sorting of the packaging waste. The licence fee to be paid for a package depends on the volume of the package and not on the type of packaging material. Some types of packaging material are easier to separate than others (steel is easier than plastics). Therefore, the license fee should be different per material, which will act as an incentive to use materials which are easy and cheap to collect and sort. Brisson (1993) concludes, however, that due to the high costs of the Dual System, 'it must be questioned whether the benefits justify these costs'.

International organizations

The developments in international organizations regarding product policy are important for the Netherlands because the Dutch environmental policy is based on international agreements (VROM, 1993b).

In the European Union (EU) an important element of the Fifth Environmental Action Programme is the principle of the shared responsibility of governments, firms and consumers to move towards sustainable development. To reach this kind of development, the use of materials and energy should be reduced (closing of the chains). Nevertheless, in the Fifth Environmental Action Programme no specific product policy is formulated. The general conditions for sustainable consumption include adapting the price of more and less environmentally-damaging products. The basis of the product policy of the EU is to encourage producers and consumers to make responsible decisions. For the consumers it is necessary that the relevant information about the environmental impacts of products is available and that the price of products is in line with their environmental impacts ('getting the prices right'). The EU stimulates producers to improve their processes and products and the availability of information they supply to consumers.

The WTO (World Trade Organization, the former General Association for Trade and Tariffs - GATT) stimulates free international trade because trade contributes to the efficient use and allocation of natural resources. However, this efficiency only holds when the costs of environmental damage are incorporated in the prices, which is usually not the case. The WTO does not have a specific product policy. It checks to ensure that national policy instruments are not contrary to the rules of the WTO, which may in fact make environmental policy making more difficult for national governments. For example, a national policy which raises import barriers (e.g. taxes) on dirty foreign products is likely to be against WTO rules (Van Beers and Van den Bergh, 1995).

Integrated environmental policies are stimulated by the OECD (Organization of Economic Co-operation and Development). The starting-point of the OECD is the Polluter Pays Principle (PPP), which states that the polluter bears the costs of pollution reduction to comply with environmental policy. In the programme of integrated life-cycle management of products and processes, the emphasis is put on information regarding products and processes obtained, for example, via life-cycle assessment (LCA) (OECD, 1994). LCA will be further discussed in Section 4.5 of Chapter 4.

In the Economic Commission of Europe (ECE) of the United Nations (UN) the emphasis of its product policy is on providing information about the environmental aspects of the chain or the product. In an Environmental Product Profile (EPP) the environmental aspects of materials and products are described in order to facilitate the communication of such information between producers and consumers.

At the Montreal Conference of the UN, it was agreed to impose a global ban on CFCs. Agreements on VOCs (volatile organic compounds) and persistent organic compounds are now in preparation.

3.7. Conclusions

This last introductory chapter has presented two approaches to environmental problems which are both used in Parts II and III of this study. The first approach is the monodisciplinary welfare approach which converts environmental impacts into money indicators. The other, multidimensional and multidisciplinary, approach focuses not only on monetary but also on physical and environmental aspects. This chapter has described the basic ingredients which may be used in studies concerned with reducing the environmental burden caused by material and product flows. These are strategies (such as recycling, substitution and technological change); policies (regulation, taxes and subsidies) and policy evaluation criteria. It should be realized, however, that various strategies may either reinforce or conflict with each other. For instance, technological change may help to facilitate the recycling of materials or reuse of products. Technological change may also affect the demand of consumers. An example of a conflicting strategy is a technological change that complicates the recyclability of materials or the repair of products. It is important to keep in mind that strategies (and policies) need to be evaluated integrally and according to various criteria. The evaluation of a strategy may have different, and even conflicting, results according to different criteria. To aggregate these results multi-criteria analysis may be used. This may also be used for comparing different strategies on the basis of multiple indices and criteria.

The concepts of chain management and M-P chains are important for analysing the possible effects of these strategies and policies. M-P chain analysis may be used to look at the effects of strategies and policies. Various criteria for evaluation may be considered, such as economic or technological criteria. These criteria depend on the model and approach that is used. In this connection various theoretical physical flow and economic models will be examined in Part II of this study. Part III will then go on to give various applications of models in which the strategies and policies discussed in this chapter are integrated.

CHAPTER 4

A SURVEY OF PHYSICAL FLOW MODELS[1]

4.1. Introduction

Different types of models have been developed to analyse the physical flows of materials and energy. These include material flow analysis, physical input-output analysis and life-cycle assessment. This chapter presents an overview of these models with their different data needs, aggregation levels, purposes and applications. The strengths and weaknesses of the models as well as their similarities and differences are discussed. For the comparison of the different methods, their particular characteristics are examined. A typology of these characteristics precedes the description and comparison of the methods. This comparison will support the selection process of choosing appropriate methods for specific problems. Finally, the relation between physical flow models and M-P chain analysis is considered.

Here, the term material flow is interpreted in a broad sense, meaning that it covers not only chemical elements, but also substances and compound materials, such as pvc. Material flows are usually measured in kilograms.

To be able to evaluate the different methods, model characteristics are compared. Section 4.2 discusses a number of important characteristics, such as the type of model and temporal aspects. Next, three main methods are discussed in turn. Section 4.3 presents material flow analysis, Section 4.4 physical input-output analysis, and Section 4.5 life-cycle assessment. Section 4.6 relates the aforementioned methods to M-P chain analysis. An evaluation and discussion is given in Section 4.7. Section 4.8 presents the main conclusions.

4.2. A Typology of Modelling Methods

Because of its complexity, the real world is not directly accessible to the human mind. Modelling is used to catch some of this complexity by focusing on certain aspects. To choose these relevant aspects, the system boundaries, the level of aggregation, and elements and relationships need to be specified. This implies that a certain part of reality can be modelled in multiple, completely different ways, depending on the goal of the study.

For researchers and policy makers it is important to know which types of model can be chosen to analyse policies and what are their respective advantages and

[1] This chapter is based on Kandelaars, Jansen and Lambert (1996).

restrictions. In addition, choices regarding the time horizon, the spatial scale, the level of aggregation, and the units of study in the model adopted, are required.[2] For each of these criteria various options exist. The choice for a certain option should preferably be in line with the options chosen for other criteria. The following set of criteria will be used as a guideline for discussing the methods in the following sections. Furthermore, this set of criteria will be used in describing the economic models of physical flows in Chapter 5, and the models applied in Chapters 6 to 10 in Part III of this study. Therefore, some criteria are broader than criteria for physical flow models alone and are directed at studying physical flows from a physical, environmental or economic perspective.

Type of model

Two types of models are distinguished: descriptive and optimization models. *Descriptive models* represent the situation at a specific point of time, or over a historical or (hypothetical) future period of time. Such models can describe the state of the environment, determine the emissions and waste that have been discharged, or assess the natural resources that are used during a certain period. When based on historical data or time series, these models may be used for forecasting purposes, provided that no major changes in the historical trend take place. If external variables follow uncertain trends, or various policies are analysed, scenarios may be developed to simulate these situations.

Optimization models are designed to maximize or minimize a single or multiple objectives. Although optimization models may describe a situation, they are different from descriptive models in the sense that the agents are optimizing or that an optimal allocation of resources is studied. An example of a single objective is to assess the optimal rate of extracting raw materials, or the optimal welfare. An example of a multiple objective model is an equilibrium model, which is based on two objective functions: for example, the profit maximization of producers and the utility maximization of consumers. The objective functions interact to obtain an equilibrium. Equilibrium models may be partial, i.e. focusing on one sector and taking the other sectors as exogenous, or general, i.e. looking at all sectors and their interactions simultaneously.

Aggregation of the model

Before modelling a certain system a level of aggregation has to be chosen. In order of decreasing aggregation the following levels can be discerned: macro-, meso- and micro-level.

(i) Macro-level (region, country): models on a macro-level are, for instance, national energy and materials use and environmental degradation.
(ii) Meso-level (product chain, sector): meso-models may be used for investigating the interactions between industrial sectors. The flow of materials and energy between the objects of study, e.g. industrial sectors, is described here. This

[2] Ayres (1978; pp. 9-15) provides some criteria for model designers some of which are used in this section.

enables the analysis of the effects of changes in one sector on other sectors.

(iii) Micro-level (product or firm): micro-models analyse firms, products or production systems.

Orientation of the models

The orientation of models describing physical flows may be directed at physical, environmental or economic aspects.

(i) Models with an economic orientation describe a national economy, an organization or a product. Quantities are generally in monetary units, and less frequently in functional units to refer to a service or a product.

(ii) Physically- or technically-oriented models describe materials and energy flows, concerning raw materials, products and waste. In such models, quantities are expressed in physical (SI-) units, e.g. kilograms and Mjoules.

(iii) Models with an environmental orientation are usually based on physical models and focus on the effects or impacts of extraction and emissions. The environmental orientation may be seen as a part of physically-oriented models. Here, this type of model is considered separately because the focus may be on units other than kilograms and Mjoules, although these units are important as an input for the other units. This orientation may account for (eco)-toxicity and other harmful effects, such as acidification and global-warming potential.

Unit studied

For the analysis of physical flows a material, a process, a product or a service may be examined.

(i) Materials-oriented studies focus on the specific materials of the flow which is described and for which alternatives and effects of policies may be analysed.

(ii) Process-oriented research investigates the improvement of existing processes and alternative options for certain processes.

(iii) Product-oriented studies focus on a specific product, and may focus on the materials and energy flows that are related to a product life cycle. Products are measured in functional units.

(iv) A service-oriented approach focuses on the desired function of the product, for instance, the transportation of x tons over y kilometres. This takes the service offered by a product as the basic criterion, rather than the product itself. Substitution between different products that are all designed to provide similar services is the basic mechanism that is studied. Services are expressed in functional units enabling the comparison of alternative products.

Temporal and spatial features

Apart from the type and focus of the model, time and space are relevant to the modelling of physical flows. Models can be either static or dynamic. To compare two states, comparative static analysis is sufficient. To analyse the transition paths between various states, dynamic models are needed.

The systems studied are usually complex systems that continuously change over time. Nevertheless, in practice often static models are applied, where time aspects are not explicitly considered. The aim of comparative static models is the comparison of different values of a particular parameter (e.g. a situation with and without a specific policy) or the comparison of a parameter at two different points in time. Dynamic models are used for studying the behaviour of systems during a continuous or discrete finite time interval. In these models time is an explicit variable that enables the inclusion of changes in technology and demand, the accumulation of materials and products, and time-lags.

The spatial scale of a model depends on the problem studied; it may be local, regional, national, international or global. For environmental modelling the issue studied is an important factor in determining the spatial level of aggregation. Some examples of environmental problems at different spatial scales are the following:

- Global: global warming, deforestation, trade of toxic waste
- International: acidification, water pollution
- National: exhaustion or landfills
- Regional: pollution of ground water
- Local: soil pollution, noise, urban air pollution

Units, performance indicators and type of variables in economic models

Physical flow models are usually formulated in physical terms (e.g. kilograms or Mjoules) or additionally in environmental terms (e.g. depletion units or global warming potential). These environmental units are constructed on the basis of physical units and provide insight into the seriousness of a problem. For the analysis of products, functional units may be used to connect the physical units with the function of a product and hence the economy.

The units in which economic models are formulated include prices, utility and quantities. Quantities, such as products and services, are usually not specified in physical or environmental terms, but in functional units. In economic models the physical quantities are not always considered explicitly, which may result in an incorrect representation of real physical states or processes. In addition, performance indicators may be used. A performance indicator is a unit that is used to summarize or aggregate (a group of) physical, economic or environmental effects, to be used in assessment and evaluation.

In addition, an important distinction is between endogenous and exogenous variables. Endogenous variables are determined within the model, while exogenous variables are determined outside the model. For example, when modelling the total consumption of a certain good, the population may be included as an exogenous variable and the consumption per individual may be endogenously determined by the price of the good.

After this typology of model aspects, various physical flow models are described, analysed and related to the concept of M-P chain analysis.

4.3. Material Flow Analysis

A material flow analysis or substance flow analysis (MFA or SFA) describes the flows of a specific material in a specific geographic area during a certain period of time. An example is the flow of cadmium in the Netherlands in 1990 (Van der Voet, 1996). Such a description allows one to analyse where the material flow may be reduced or changed. It provides insight on the material flowing and accumulating through the economy.

MFA describes the total flow of a specific material through the economy, thus including all products which contain or use that material. MFA is based on the material balance (MB) principle (see Section 2.4). This allows one to trace missing flows of materials and to identify and predict environmental problems. The MFA method considers one or more material types. If more material types are considered, they are described separately and independently of the service they provide. Thus, interactions between the different flows are not taken into account, which implies that substitution and complementarity between materials cannot be observed.

MFA is a tool to provide insight into the material flows in an economy, which may support the decision making on environmental policies related to materials. Because MFAs are usually descriptions, they are inappropriate to analyse societal, economic or behaviourial mechanisms.

Static MFA studies describe material flows in one period of time. Dynamic MFA studies analyse the (historical) material flows over a certain time period, and allow for changes in the extraction or use of materials. MFA studies are usually carried out on a national or regional level, although they can also be performed on a more detailed level, for instance, on a firm level. MFA studies rarely incorporate specific environmental indicators, but they provide information to derive such indicators.

Figure 4.1 gives an illustration of an MFA study in which the flows through the economy and the environment are described. The accumulation of materials in the economy and environment may give rise to environmental problems in the future (Guinée *et al.*, 1998).

There are numerous recent studies on the flows of one or more materials in a specific geographical area in a certain time period.[3] On the basis of MFA, models may be built to analyse or estimate flows when certain policies are imposed (i.e. a scenario analysis). For the cadmium and nitrogen flow in the EU a descriptive model is used to estimate changes in material flows if the scheduled policies are implemented (Van der Voet, 1996).

MFA has its origins in (economic) I-O modelling in which the (value or economic) flows through the economy are described, and according to Van der Voet (1996; p. 7) another origin of MFA is the detailed bookkeeping of agricultural materials. Physical and economic (including environmental) I-O modelling will be discussed in Sections 4.4 and 5.4 respectively. MFA and I-O analysis describe the

[3] *Heavy metals* have been studied for several regions (Anderberg *et al.*, 1989); Ayres *et al.*, 1989; Bergbäck *et al.*, 1992; Stigliani and Anderberg, 1992 and 1994; Van der Voet *et al.*, 1994a; Gilbert and Feenstra, 1994; Gorter, 1994; and, Ayres, 1994). Various MFAs have been done for different types of *chemicals*, e.g. nitrogen and sulphur (Ayres and Norberg-Bohm, 1992a and 1992b; Husar, 1994), and chromium and lead (Lohm *et al.*, 1994). MFAs on other materials or substances have been performed by Brunner *et al.* (1994) for a Swiss region, and Ayres *et al.* (1994) for carbon monoxide and methane in the United States. Some *dynamic material balance* studies have been done, e.g. for fly ash (Olsthoorn *et al.*, 1991) and micropollutants (Olsthoorn, 1991).

material flows in the economy. MFA usually describes a material flow, or more material flows separately. Therefore, MFA may be considered as a physical I-O analysis for one material. In principle, MFAs for various material flows may be aggregated to obtain a physical I-O analysis for various materials. Thus, methodologically speaking, MFA and physical I-O analysis are the same, implying that the drawbacks of both methods are also the same. These will be discussed in the next section on physical I-O analysis.

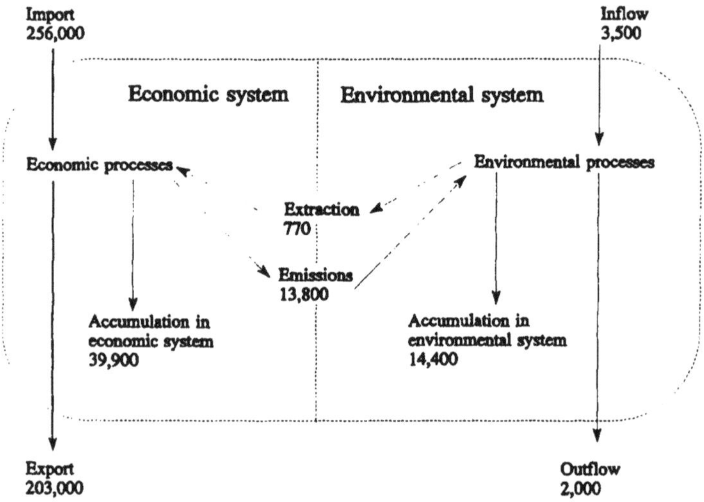

Figure 4.1. MFA study on zinc flows in the Netherlands in 1990 (in tonnes) (adapted from Annema et al., 1995).

4.4. Physical Input-Output Analysis

Input-output (I-O) analysis was introduced by Leontief in 1941 and has been widely applied and further developed and modified (see also Leontief, 1966).[4] I-O analysis is a quantitative macroeconomic tool that is based on National Accounts. The purpose of the I-O framework is to analyse the interdependence of industries in the economy and therefore it is also referred to as 'interindustry analysis' (Miller and Blair, 1985). An economic I-O table describes in monetary units the mutual exchange of goods and services between the different sectors of industry. Analogous

[4] A method strongly related to I-O analysis is activity analysis. This method is described in Section 5.4 of Chapter 5.

to monetary I-O models, physical models have been elaborated based on the same philosophy of bookkeeping transactions between sectors. Whereas accounting in monetary units allows for aggregation of different types of products and services, accounting in physical units can only deal with disaggregated flows of a single, homogeneous product or material (Miller and Blair, 1985). Physical I-O models use the material balance principle, stating that physical inputs equal outputs. Monetary I-O models do not use this principle.

This section presents a physical I-O table in which the physical interactions, i.e. inputs and outputs, of various sectors are systematically described. These may be used for analysing environmental issues related to the use of materials or products. A basic monetary I-O table with resources and emissions is described here. Later on, Section 5.4 of Chapter 5 discusses augmented (economic) I-O tables and integrated environmental-economic I-O tables.

I-O tables are in homogeneous units. It is assumed that every firm produces one output. To make a physical I-O table, a firm is assigned to an industrial sector based on the output it produces. However, firms in one sector and even one firm may produce multiple types of products. Furthermore, when producing an output other products, co-products, by-products and waste are produced (Heijungs, 1997). The multi-input/multi-output approach to material and energy flows specifically deals with the various input and output flows of processes (Jansen and Lambert, 1996). Thus an I-O table has industrial sectors and the corresponding outputs (groups of products) as the row and column elements, respectively.

In I-O models the ratios between inputs and outputs of a sector are constant (fixed technological coefficients). This type of relationship, known as a 'Leontief production function', assumes constant returns to scale and full complementarity between inputs. This makes it impossible to incorporate substitution processes or technological change in the basic I-O model.

An I-O table (see Table 4.1) consists of the intermediate deliveries between economic sectors (AX, with A the matrix of input coefficients in the economic sectors, and X the vector of the output of the economic sectors), the deliveries from the economic sectors to the final demand (Y), and the use of primary physical inputs to the economic sectors (LX, with L the matrix of input coefficients from primary inputs to the economic sectors). Let vector E denote the total of required primary inputs. The vectors X, E and Y are in physical terms. The first three rows of Table 4.1 present a standard I-O table which leads to two equations: (i) $AX+Y=X$, that is, demand equals supply; and, (ii) $E=LX$, that is, the total required primary inputs are proportional to the total supply. The first equation may be rewritten, so that the total supply is a function of the final demand, $X=(I-A)^{-1}Y$. The term $(I-A)^{-1}$ is called the Leontief inverse and represents the cumulative direct and indirect use of intermediate goods per unit of final good. The second equation may be rewritten with the use of the first equation as $E=LX=L(I-A)^{-1}Y$. This implies that when the total output (X) changes the primary inputs (E) change, and when the final demand (Y) changes the pollution changes via the indirect or cumulated matrix of coefficients $(I-A)^{-1}$. The last two rows present the resource use of, and the pollution generation by, the economic sectors. The resources (R) and pollution (emissions) (P) may be calculated via the direct impact coefficients (C for resources and D for pollution): $R=CX=C(I-A)^{-1}Y$ and $P=DX=D(I-A)^{-1}Y$. In Section 5.4 of Chapter 5 an abatement sector is added to the I-O table.

These rows show the interdependencies between the economic sectors, the final demand, the resources and the emissions. The table may be used to measure the effects of a change in the final or intermediate demand (X or Y) or the technical coefficients (L and A) on all sectors. A strength of I-O tables is that not only can the direct effects be calculated but also the indirect ones. The indirect effects may be calculated from the interactions between the sectors (i.e. the economic structure). For example, a decrease in the final demand of sector i affects the inputs of sector i. Then, if sector j supplies an input to sector i, the demand for the output of sector j is affected. Moreover, also the inputs required for sector j change, which may ultimately again affect sector i, etc. Therefore, a change in the output of one sector may affect other sectors.

Optimization models may be developed based on descriptive I-O models. For instance, the value added may be optimized given the I-O production structure. Alternatively, resource use or emissions may be minimized given a restriction on demand.

Table 4.1. An I-O table with resources and emissions.

	Economic sectors	Final demand	Total input	Total output
Economic sectors	AX	Y		X
Primary inputs	LX		E	
Resources	CX		R	
Pollution	DX			P

4.5. Life-Cycle Assessment

Although several definitions of life-cycle assessment (LCA) have been proposed, the ISO-14000 definition has been set as a worldwide standard (ISO, 1995): Life-cycle assessment is a systematic set of procedures for compiling and examining the inputs and outputs of materials and energy and the associated environmental impacts directly attributable to the functioning of a product or service system throughout its life cycle.[5] LCA is intended for comparative use, i.e. the results of LCA studies have a comparative significance rather than providing absolute values on the environmental impact related to a definite product. Therefore, in LCA two or more alternatives are compared. These may be existing products, or potential, new or improved products.

Essential to LCA is the 'cradle to grave' approach, taking into account material

[5] In principle, the concept of LCA can be extended to include allocation and economic processes. Here, the term 'M-P chain analysis' is used to emphasize the difference with LCA as defined by ISO (1995) and Guinée (1995).

and energy flows from extraction up to waste treatment. The basis of the description of physical flows related to a certain product is the input and output in economic activities and environmental impacts. The environmental effects of energy use, capital goods and by-products are included. Although various economic, social and safety aspects of a product during its entire life cycle may be studied, the concept of LCA is usually confined to a quantitative analysis of environmental aspects.

LCA is usually carried out in five phases (Guinée *et al.* 1993a and 1993b; ISO, 1995; Berg *et al.*, 1995). These phases are: (1) goal definition; (2) inventory of environmental inputs and outputs; (3) conversion of inputs and outputs to environmental impacts; (4) (optional) evaluation, i.e. comparison of environmental impacts to some standard; and, (5) improvement analysis. LCA studies are specifically performed to examine the environmental aspects of a product.[6] The conversion of inputs and outputs to environmental impacts lead to the 'attribution problem' which is "the question which environmental problems are to be attributed to which economic activity", or more specifically to which material or product (Heijungs, 1997, p. 4).

LCA studies generally do not cover complete material balances, because for practical reasons only the environmentally most harmful flows are considered. However, a complete inventory of materials and energy use is recommended because it may lead to valuable options for saving raw materials and energy. In comparison to MFA and physical I-O analysis, LCA is not related to a time period in which the flows are measured. Another difference between MFA and LCA is that MFA studies a material in physical units, while LCA studies a product measured in a functional unit (i.e. a unit of a product). Although LCA is a potentially useful tool to perform a complete inventory of physical flows, this is often not accomplished due to data restrictions. LCA software supports the storage of required data, and a proper presentation of the results.[7]

The concept 'Produktlinienanalyse' makes a catalogue of criteria to which the various products may be ranked (Osnowski and Rubik, 1987). As in an LCA study, in a 'Produktlinienanalyse' the environmental aspects of a product are assessed from 'cradle to grave'. In addition to LCA, 'Produktlinienanalyse' also lists private and external costs, and social aspects, such as unequal wages and working conditions, that can be attributed to the product. However, no market mechanisms, economic processes or behaviour are included is this concept. Additionally, determining the external costs and social aspects empirically leads to fundamental methodological problems (Osnowski and Rubik, 1987, p. 84).

Physical flows related to a product have various environmental impacts, such as acidification. In order to make these impacts comparable equivalence factors are

---

[6] Examples of LCAs concern: building materials, energy carriers (Frischknecht, 1994), plastics, petrochemicals, paints and varnishes, margarine, hairsprays, detergents, etc. (Guinée, 1995), packaging materials, faceplates, computer cases, automotive parts (Snowdon, 1994; Brinkley, 1994; Eyerer, 1993), automotive vehicles (Schuckert, 1993).

[7] An example of a software tool for LCA of products is SIMAPRO (Guinée *et al.*, 1991). SIMAKOZA has been developed specifically for window frames (Guinée *et al.*, 1992).

used in the evaluation phase of the study. With weighted aggregation of the various environmental impacts, LCA results may be presented in terms of one or more environmental performance indicators (EPIs). EPIs are strongly aggregated characteristics that reflect the performance of a complex system. The basis of an EPI may be an LCA, but also other environmental data may be presented in an EPI.

EPIs can be aggregated into a single characteristic, for instance, the Material Intensity Per Service unit (MIPS) (Schmidt-Bleek, 1993).[8] An example of an EPI in which a limited number of environmental impacts are considered is the Ecological Footprint (Wackernagel and Rees, 1996).[9] EPIs are applied to communication within environmental management systems and to external reporting. Like LCA, EPIs are only suitable for comparative purposes and do not provide any reliable absolute figure. Because most EPIs strongly depend on the applied weight factors, these should be well documented in reporting to guarantee adequate exchange of information and appropriate comparison of data. A criticism of EPIs is that they aggregate various non-comparable impacts and that the resulting indicator no longer provides any clear interpretation. EPIs may be useful in communication directed to policy makers and stakeholders of companies, where recognizability and comparability are the principal requirements. For scientists these requirements are also valuable, but they would like to have less aggregated, more objective and more accurate data on the environmental impacts.

4.6. Physical Flow Analysis and M-P Chain Analysis

The driving force for the consumption of products is the desire for services. Therefore, materials, products and services should be studied together and simultaneously. The concept of an M-P chain includes an economic structure of connected material and product flows (see Section 1.2 in Chapter 1). With the concept of an M-P chain various analyses can be performed. LCA studies an M-P chain because LCA examines the economic structure of material and product flows. MFA, however, does not study an M-P chain, because it does not include products. In this study 'M-P chain analysis' studies the allocation and economic processes of an M-P chain. This definition does not include LCA, MFA and I-O analysis,

---

[8] MIPS refers to a single aggregated indicator representing the total direct and indirect use of materials related to a unit of service, expressed in units of mass (kilograms), ignoring the environmental characteristics of the different materials that are involved. This technique is primarily designed for providing information (Fresenius Environmental Bulletin, 1993; Hinterberger *et al.*, 1995). Bringezu (1993) links LCA to MIPS to quickly screen the materials that are needed for a certain product or functional unit.

[9] The concept of 'Ecological Footprint' is an aggregate indicator. Selected flows of materials and energy are calculated and converted into the corresponding hypothetical land area needed to support these flows. This indicator may be compared with the land area that is actually used in a region or country. Criticisms of this indicator are that only a number of flows are included; the aggregation of the flows is unclear or arbitrary; no differentiation is made between sustainable and non-sustainable land use; a region may always be chosen in such a way that it is unsustainable (e.g. a city).

because these methods do not include allocation of materials and products, and economic processes. M-P chain analysis uses elements from MFA, physical I-O analysis and LCA, and combines those with an economic analysis.

An economic analysis of M-P chains allows the study of, for instance, optimization, market equilibrium, policy analysis and scenarios for future development. M-P chain analysis may give insights into the reduction of the environmental burden caused by the demand for a service. Aspects that can be a part of an economic analysis of M-P chains, and the differences with the methods discussed in Sections 4.3 to 4.5, are listed below:

- *Economic processes*, such as prices and costs of materials and products, market equilibrium, allocation, production functions, are explicitly modelled. The *behaviour* of consumers and producers is included because these influence the use of materials and products. Economic, monetary or behaviourial aspects are generally not included in LCA and MFA studies. In I-O studies these aspects are only incorporated on a highly aggregated level.
- *Recycling* of materials and reuse of products is fully included. In LCA, recycling and reuse may be included too. In MFA, on the other hand, recycling is taken into account only for materials.
- M-P chain analysis incorporates the *substitution* of different materials or products, and of materials and other inputs. MFA and I-O studies, in contrast, do not include substitution. To compare products, LCA studies consider substitution of products. Substitution at a material level alters the description of the flows in LCA, MFA and I-O analysis.[10]
- Both long-range changes in time, e.g. technological developments, and short-range changes, e.g. changes in demand, are considered. In other methods *dynamic aspects* are generally excluded. Recently, however, there are some MFA and LCA studies in which dynamic aspects are included (Gilbert and Feenstra, 1994; Moll, 1993).
- The goal of M-P chain analysis is to analyse the effects of various instruments or *policies* on the physical flows and on environmental and economic indicators. The goal of LCA is to compare various products, while that of MFA and physical I-O analysis is to describe the physical flows.

Ayres (1995) argues that the integration of process analysis with conventional sectoral models (e.g. I-O models) provides important improvements for both types of models. M-P chain analysis includes some of the integration aspects of those models:

[10] The equations in LCA, MFA and physical I-O analysis describe the material flows as fixed technological coefficients, as in an input-output framework, i.e. $Y=\alpha X$ with α a fixed (technological) coefficient between one input X and one output Y. Such a set of (homogenous) linear equations may be written as a matrix. In M-P chain analysis the equations may be as follows: $Y_1=f(X_1,X_2)$ with the output Y_1 as a function f of inputs X_1 and X_2; or, $g(Y_1,Y_2,X_1,X_2)=0$ in which a function g includes multiple variables, which may be interpreted as outputs Y_1 and Y_2 (e.g. positive variables) and inputs X_1 and X_2 (e.g. negative variables). The functions f and g may be non-linear, and include multiple inputs and outputs. The equations of an M-P chain analysis cannot be written as a matrix, because the technological coefficients may vary and be non-linear.

namely, material balance conditions; substitution and multiple inputs and outputs which form part of the production function; recycling; and, non-linearities. Furthermore, behaviourial and economic aspects are incorporated.

A specific type of model related to M-P chain analysis is the 'materials-process-product model' (Ayres, 1972 and 1978). This is a model to facilitate quantitative analysis of material flows through the economy, where alternative technological possibilities exist (Ayres, 1978, pp. 154-165). In this model an industry maximizes its profit (or minimizes its costs) for the production of a product under a limited set of fixed technological options. The material balance conditions hold for each process (or activity) in the production process. The result is a ranking of alternative production options. In a more advanced model 'process loops' such as recycling loops may be included. The basis of a materials-process-product model is an M-P chain, in the sense that an economic structure of material and product flows is studied (see Section 1.2 in Chapter 1). A materials-process-product model may be seen as basic type of M-P chain analysis (in the narrow definition, see Section 1.2 of Chapter 1), because allocation and economic processes are included. However, in a materials-process-product model only one industry is analysed, and it does not include: consumers' choice of a product to fulfil a certain service; a market equilibrium; explicit modelling of economic processes; endogenous behaviour of agents; optimization on a social or chain level (only on an industry level); and, (policy) scenarios for environmental development.

Depending on the goal of the M-P chain analysis, a certain part of the interlinked material and product flows (i.e. a truncated M-P chain, see Figure 1.1) is chosen. In principle, both descriptive and optimization models can be used (see Section 4.2). An M-P chain analysis should be confined to a certain geographical area, which can be a region or a country. The environmental aspects (see, Section 4.2) can be dealt with separately or simultaneously. For example, to reduce the consumption of materials or the disposal of waste material an M-P chain analysis can result in data on the effects of a certain policy on the chain. M-P chain analysis can be performed with static, comparative static or dynamic models. The choice of the time horizon in dynamic models depends on the goal of the analysis. Services and products are the central issue of M-P chains. Thus, the aggregation level of this method is the micro-level. When studying product groups - for example, when using I-O models - the analysis may be performed on a meso-level (see Chapter 10).

M-P chain analysis is, like LCA and MFA, limited by data availability and unpredictable future flows. Moreover, certain criteria are needed to truncate all the related material and product flows, and assumptions need to be made on the uncertainty of prices and the impact of policies on consumer and producer behaviour.

4.7. Evaluation and Discussion of Methods and Characteristics

In this section the methods described in Section 4.3 to 4.6 will be assigned to the characteristics discussed in Section 4.2, in order to indicate the differences between various methods. Table 4.2 presents some of these characteristics for the various methods. The strengths and weaknesses of the respective methods in relation to their different aspects of application are evaluated.

A physical flow may be related to other physical flows: for instance, because products are made of various materials. Physical flows are embedded in an economic system that requires a certain number of products. Therefore, for modelling physical flows in an economic system two types of interfaces are important: first, interfaces between a number of physical flow models, in order to build models of more extended systems, e.g. product chains or interactions between different chains; and, second, interfaces between physical flows and the economic system.

Table 4.2. Overview of some characteristics of the modelling methods.

Method	Type	Aggregation level	Orientation	Unit studied
MFA	Descriptive	National or regional level (macro-)	Physical	Materials (kilograms)
Physical I-O analysis	Descriptive/ optimization	National or regional level (macro-)	Physical	Materials (kilograms)
LCA	Descriptive	Product chain (meso-)	Physical and environmental	Products (functional unit)
M-P chain analysis	Descriptive/ optimization	Product chain (meso-)	Physical, economic and environmental	Service or products (functional unit)

Type of model

The different methods of modelling have as a common basis a description of the physical flows studied. The (historical) description may be extended to a model in which scenarios may be analysed. Most MFA studies are descriptions. LCA studies, also mainly descriptive, are usually applied to compare different products or processes. For I-O analysis and M-P chain analysis a description may be linked to an optimization model, or for M-P chain analysis to an equilibrium model.

Aggregation of the model

MFA and I-O models are mainly performed on a sectoral level to study the flow of materials between sectors. The flows are generally studied for the whole economy. The aggregation level may be seen as a mixture of macro- and meso-levels. LCA and M-P chain analysis deal with products and the whole production chain on a more detailed level. This may be seen as a meso-micro-level of aggregation.

Orientation of the models and unit studied

MFA, LCA, physical I-O models and M-P chain analysis are based on the description of inputs and outputs in economic activities and environmental systems. MFA studies are physically oriented with a focus on environmental issues related to the materials described. LCA studies are physically orientated too, but the unit studied is a product or a service. Physical I-O analysis looks at the materials and products of sectors, which are mainly linked to economic sectors. M-P chain analysis studies the physical and economic aspects of a service or a product simultaneously. MFA, LCA and M-P chain analysis are also oriented towards environmental aspects. In physical I-O analysis, depletion and pollution aspects measured in kilograms may be included.

Temporal and spatial features
MFA, physical I-O analysis, LCA and M-P chain analysis are usually performed statically. In practice, MFA and physical I-O analysis may be dynamic with the use of time-series data, but this is rarely done due to a lack of appropriate data. I-O tables may be made for several years separately with different technical coefficients. The data on which LCA is based are independent of time. However, time-series data may be included by distinguishing products over time: for example, as a result of a change in production technology. M-P chain analysis may include time-series data in order to distinguish products and materials. For all types of models, the impact of inaccurate input data should be evaluated by sensitivity analysis. All types of modelling are applicable to scenario analysis and other forecasting or evaluation techniques.

LCA and M-P chain analysis are usually confined to products or services. MFA and I-O analyses, however, refer to existing, sometimes extensive, geographic regions. Most MFA and I-O analyses are done on a national scale.

Units, performance indicators and types of variables
Of the four quantitative model types discussed in this chapter, MFA analyses the physical flow of materials in kilograms (or tonnes) through the economy from and to various environmental compartments, such as air, ground and water. Thus, there is a link between environmental or economic units and the physical flow, although not much attention has been given to quantifying these environmental or economic units. Physical I-O models analyse physical flows on a macro-level which results in studying an aggregate physical flow measured in kilograms or Mjoules. Physical and monetary I-O models may be linked (see Chapter 5). LCA focuses on a product, which is an economic unit, and measures the environmental impact in various units (energy use, global-warming potential, acidification) for the whole life cycle of the product. In LCA the physical and environmental aspects are considered, but not the monetary units. Monetary, physical and environmental aspects are considered together in M-P chain analysis. Not all physical flows are taken into account as in MFA and I-O modelling, but only a limited number of flows which seem to be the most important physically, environmentally or economically. Performance indicators may be used to aggregate various environmental variables to compare the environmental impact of, for example, different kinds of products.

4.8. Conclusions

From this survey of the most commonly applied physical flow modelling methods and M-P chain analysis, and the typology of their relevant characteristics, the following main conclusions are drawn.

In current applications there is an emphasis on descriptive and static representation of physical flows. Recently, though, dynamic aspects are beginning to be considered in all four methods described. Thus, dynamic issues may be included in all models, but the data required is often lacking. For both static and dynamic models, the need for accuracy and completeness of data is an issue that continually deserves attention.

In many of the studies described in the literature, the level of aggregation is relatively high. This is mainly due to difficulties of data acquisition, and lack of knowledge of the exact flows of materials and products. LCA and M-P chain analysis study physical flows on a product or service level which makes it possible to examine the consumption side of product flows in detail, but without considering the heterogeneity or the decisions of individual producers or consumers.

Physical flow models are useful in analyses that are characterized by a broad spectrum of temporal and spatial scales, level of aggregation and orientation. As has been pointed out, however, different modelling techniques may be suitable for different purposes. This will also be shown in Part III of this study in which different models and their applications will be presented.

Substitution, recycling and reuse are important issues for the analysis of the environmental impact of materials and product use and therefore should be considered fully. If recycling or reuse occurs it needs to be included in LCA and in M-P chain analysis. In MFA and physical I-O analysis recycling may be included. Substitution is included on the level of products in both LCA and M-P chain analysis. Substitution of materials is only analysed in M-P chain analysis. MFA and I-O analysis do not include substitution.

Physical flows result from the demand by producers and consumers for materials and products. This demand depends on economic factors such as prices and preferences. In MFA, I-O models and LCA these economic factors are not (adequately) taken into account. The physical flows are described without looking at their economic basis. From a physical and environmental perspective M-P chain analysis consists of elements of MFA, physical I-O analysis and LCA, but adds substitution, recycling and allocation processes.

Chapter 5 discusses various environmental economic models in which material flows may be included. Chapters 6 to 10 present various models and applications of M-P chain analysis.

CHAPTER 5

A SURVEY OF MATERIAL FLOWS IN ECONOMIC MODELS

5.1. Introduction

Historically, in economics, the flows in and out of production and consumption are considered mainly in value, utility or money terms, while no or little attention is given to the flows in material or energy units. Recently, however, the interest in studying physical flows in economic processes has begun to grow, mainly due to their associated environmental problems.

There are a number of economic model types that address environmental problems. In this chapter it is examined whether these can be applied to the analysis of material flows. It is important to know which types of models may be chosen to deal with particular research questions that address economic, environmental or physical aspects of material flows. Five main categories of models are considered. It is not attempted to give a complete overview of all aspects of all models, nor to give a complete account of all applications. Instead, the main research questions related to the economics of material flows are discussed for each model type, while examples illustrate how material flows may be integrated in the model types considered.

The organization of the chapter is as follows. First, two types of models based on neoclassical economics are examined. Section 5.2 discusses economic models of natural resources. Section 5.3 presents pollution models dealing with the optimization of pollution activities in a neoclassical framework with externalities. Section 5.4 addresses various types of environmental input-output models, allowing the direct and indirect use of materials to be calculated. Macroeconomic models are discussed in Section 5.5. In the context of the use of materials they are mainly used for the description and prediction of this kind of use on a national level. Section 5.6 describes dynamic and evolutionary models dealing with the impact on material flows of evolutionary, technological and structural changes in the economy. Each of the Sections 5.2 to 5.6 includes a general introduction to the model type and its relation to material flows followed by examples, particularly sub-classes of models, and an overview of some core characteristics of the model type. Section 5.7 draws conclusions and discusses the applied studies of Part III of this study against the background of the theoretical models in this chapter.

5.2. Economic Models of Natural Resources

Economic models of natural resources mainly deal with the problem of the intertemporal allocation of resources and, related to this, missing markets for many resources (Neher, 1990). This intertemporal, 'now-or-later', setting requires a dynamic analysis.

A common distinction is between renewable resources, such as fisheries and forestry, and non-renewable resources, such as minerals and fossil fuels. Renewable resources are capable of regeneration over time as long as their surroundings remain favourable and the stock does not become so small that it might go to zero (e.g. extinction in fisheries). On the other hand, non-renewable resources do not regenerate, or at least not at a significant rate on a human time scale, so that the use of a non-renewable resource will reduce the stock forever.[1]

The main approach of these economic models is the following. Note that this description does not pay attention to nuances. Economic models of natural resources focus on optimizing a certain goal, for example social welfare or the sum of discounted private profit. They determine the socially optimal extraction path for non-renewable resources, the maximum production, the sustainable optimum population of fish or animals, the optimal sustainable catch, the optimal depletion rate, or the maximum present value of resources. The technique that is used for the intertemporal analysis of natural resources is dynamic optimization or optimal control in which the control variables may be policy or technical variables. The Hotelling rule can be used to derive the optimal depletion rate for non-renewable resources or an optimal harvesting rate for renewable resources. According to that rule, this rate of depletion or harvesting rate must be equivalent to the increase in the price rate of the resources plus the interest rate. In other words, the present, discounted value of the resource should be the same at all dates (Hotelling, 1931).

Economic models of natural resources and material flows

The research questions addressed in economic models of natural resources are also useful in the context of material flows. Besides the usual division between non-renewable and renewable natural resources, the distinction between recyclable and non-recyclable materials is relevant for material flows studies. The optimal allocation of natural resources and associated material flows, with or without recycling and regeneration, may then be studied with economic models of natural resources. Important aspects related to material flows that may be taken into account by economic models of natural resources are recycling, substitution, technological progress and the material balance principle. These topics will now be discussed in relation to particular economic models of natural resources.

Recycling may contribute to resource conservation, depending on the percentage of materials that economically and technically may be recycled. Weinstein and Zeckhauser (1974) offer a theoretical study of the optimal consumption pattern of depletable resources with recycling. Lusky (1975, 1976) includes recycling in a dynamic optimization model of a natural resource cycle, i.e. a material flow with recycling and pollution. It is concluded that only with governmental interference can

[1] Both renewable and non-renewable resources are depletable and exhaustible: renewable resources become exhausted when the population size or the stock becomes too small; and, non-renewable resources are exhausted when the stock becomes zero.

an optimal solution be reached. Hoel (1978) studies the connection between resource extraction and recycling in relation to the environmental costs in a dynamic optimization model. Max and Lehman (1988) study the optimal uses of a non-industrial private forest. A landowner optimizes the utility, given the recreation and income function minus the tax, over two periods, which results in an optimal harvest rate. This model illustrates the 'now-or-later' problem in relation to a stock of materials that may have economic functions at different times.

Substitution of materials by other materials ('direct' or 'inter-material' substitution) is an option for the conservation of natural resources which is typically not studied in economic models of natural resources. A possible explanation is that these models mostly focus on one material or sector. The substitution of materials by labour ('indirect' substitution) has also been studied in the context of technological change which may be considered in the dynamic economic models of natural resources. These models mainly focus on the extraction of resources and not on the pollution side. Therefore, the material balance condition is not explicitly taken into account in most models, but, of course, in the models which also deal with pollution the material balance condition may easily be incorporated.

Production function and material flows

A production function is important for all model types discussed in this chapter. For modelling material flows, the production function will evidently include natural resources as an input apart from labour and capital. The production function is not directly related to the choice of extraction or the allocation of resources in economic models of natural resources, but this function may be important: for instance, to determine the costs of extraction or production. The input of materials in a production function needs to be equal to the output of materials according to the material balance condition (see Section 2.4). For a production function in which materials (together with labour and capital) are converted into a product, this material balance condition may be used to keep track of the material contents of products and the waste material resulting from disposal of the products. This is usually not done in economic models of natural resources.

The standard formal production function (F) with three inputs, capital (k), labour (l) and natural resources (r) and with output (x) is $x=F(k,l,r)$. If there is no resource input, the output will equal 0, $F(k,l,0)=0$, and if there are positive inputs $(k,l,r>0)$, then the output will be positive too, $F(k,l,r)>0$. In a dynamic model technological progress can be included by adding time (t): $x=F(k,l,r,t)$, or, for example, a state of technology at time t, $A(t)$, which gives the production function $x=F(k,l,r,A(t))$. With a change in technology the amount of inputs required to make one output may be reduced. This allows less natural resources to be used, prolonging the time until the exhaustion of the stock of resources.

In the standard production function, material balance conditions are not considered, but they may be taken into account by looking at the material contents of a product. For example, the production function (F) of product X and production inputs capital (k) and materials (r) is $X=F(k,r)$. The material contents of the product is then r. After consumption the product (X) may be transformed into waste material (w) by a transformation function G in which the material contents of X is transformed into waste material, w=G(X). The material balance condition of this process is that the waste material (w) equals the material inputs in the production function (r), w=r. Other examples of material balances in production functions are given in Gross and Veendorp (1990), and Van den Bergh and Nijkamp (1994). In dynamic models the material balance conditions may result in depletion or pollution problems that would not have been observed without these conditions (for example, in growth models).

Policy instruments
Various policies can be imposed on resources and each of these has its advantages and disadvantages (see Chapter 3). For non-renewable resources a revenue tax may be imposed which will increase the time left before complete exhaustion. Some examples of policies which can be imposed on renewable resources are the following:

- A tax per unit of the natural resource caught, extracted or harvested.
- A transferable or marketable catch or extraction permit. The total number of permits is determined by the total allowable catch or extraction.
- Technical restrictions: for example, for fisheries a limit on the number of boats and the number or size of the nets.
- Quantity restrictions on catch or extraction.

An example an economic model of natural resources
To illustrate the issue of optimal extraction and the social welfare function, a simple two-period model for a fixed stock of a non-renewable resource is presented. The model includes recycling the resource in the first period allowing the recycled resource to be used in the second period. This model is an extension of a model by Perman *et al.* (1996, Chapter 6). The two-period welfare function, W, is:

$$W = U_0 + \frac{U_1}{1+\rho}$$

with utility at time t, U_t, and ρ as the social utility discount rate. The utility equals the net social benefit at time t, i.e. the revenue (the quantity extracted, Q_t, times the price, P_t) minus the costs (the costs times the quantity extracted, $c*Q_t$). Note that the quantity extracted equals the quantity demanded. The price depends on the quantity demanded (i.e. inverse demand function), P_t=a-bQ_t. Then the gross social benefit becomes:

$$\int_0^{Q_t} (a-bQ)\delta Q = aQ_t - \frac{b}{2}Q_t^2$$

The objective function is the optimization of the net social benefit, i.e. gross social benefit minus cost of extraction (cQ_t, for t=1,2) and of recycling (dR_0), over the two periods:[2]

$$\max_{Q_0,Q_1} W = aQ_0-\frac{b}{2}Q_0^2-cQ_0-dR_0 + \frac{aQ_1-\frac{b}{2}Q_1^2-cQ_1}{1+\rho}$$

subject to the recycling function $R_0=f(Q_0)$, and the total quantity extracted equals the fixed initial stock of non-renewable resources, S^*.

$$Q_0+Q_1+R_0 = S^*$$

Solving this optimization problem gives the following condition for optimal extraction:

$$(P_0-c-\frac{df}{dQ_0})(1+\rho) = (P_1-c)(1-\frac{df}{dQ_0})$$

The prices which give the welfare optimization, P_0 and P_1, can then be determined.

Under a revenue tax policy of α per unit of resource sold, i.e. revenue per unit equals $P_t(1-\alpha)$-c, the optimal condition becomes:

$$(P_0-\frac{c}{1-\alpha}-\frac{df}{dQ_0})(1+\rho) = (P_1-\frac{c}{1-\alpha})(1-\frac{df}{dQ_0})$$

This tax prolongs the time left before the exhaustion of the stock. This model shows how natural resources can be included in a neoclassical welfare function and how a policy may affect the welfare function.

Overlapping generations and discount rates

Models with overlapping generations allow one to deal with various generations that exist at each point in time and to take into account generations which are not yet born (Blanchard and Fisher, 1993). In economic models of natural resources the natural resources of future generations may be taken into account to obtain the optimal level of extraction. Howarth and Norgaard (1995) state that for an intertemporal allocation of resources and intergenerational equity a sequence of overlapping generations is needed. Howarth and Norgaard (1993 and 1995) discuss the optimal level of intergenerational transfers, e.g. transfers between parents and children, in a model with overlapping generations and an exhaustible resource. The transfer of assets, such as natural resources, from the present to the future generations may encourage sustainability because one generation considers the following generations in their utility function. The discount rate is best seen as a market price, despite many (ethical) controversies on this topic (Daly and Cobb, 1994). The

[2] Since the model is a two period model there is no recycling in the second period because recycling is costly and the recycled resources will not be used any more.

example above may be seen as a two-generations model with discount rate ρ. It is not a model of overlapping generations because the utility functions U_1 and U_2 do not affect one another.

Aspects of economic models of natural resources
Some of the characteristics discussed in Chapter 4 are listed here to give an overview which may also be used to compare economic models of natural resources with the other model types in this chapter. These characteristics should be considered as rough indications and not as hard facts.

Table 5.1. Dominant characteristics of economic models of natural resources.

Problem to solve	Optimal allocation of natural resources over time
Type of model	Optimization, partial
Time aspect	Dynamic
Spatial scale	Dependent on (the stock of) natural resources studied
Aggregation level	Sector (fisheries, forestry, mining) or firm
Units	Prices, quantities and utility

5.3. Pollution Models

Whereas the models in the previous section mainly discussed the intertemporal allocation of resources, a considerable number of environmental economic models focus on valuation and regulation of pollution (Cropper and Oates, 1992). In this section pollution models are discussed. These models consider the regulation of polluting economic activities by imposing policy instruments. This section does not address the topic of the valuation of the environment and resources.

The theoretical basis of pollution models is neoclassical welfare economics and in particular the concept or externalities. The types of models which are used in pollution modelling are optimization and equilibrium models. In optimization models profit, social welfare, costs or environmental damage may be optimized under a set of constraints, such as a production function, a budget condition or a material balance. An 'environmental manager' may optimize with a set of physical, environmental or economic restrictions. In Chapter 6 an optimization model will be presented in which an environmental manager optimizes the total costs to fulfil the demand, when considering the M-P chain for various products. Optimization models may be either static or dynamic. For both economic models of natural resources and pollution models the same dynamic optimization techniques are used.

Equilibrium models are characterized by utility and profit maximization and market clearing. In equilibrium models the equilibrium price results from the optimizing behaviour of producers (maximization of profit under technical constraints) and consumers (maximization of utility under budget constraints). The resulting equilibrium price clears the markets of goods and production factors.

Equilibrium models can be either partial or general. Partial equilibrium models focus on one market or one sector, while all other prices and markets are assumed to be fixed. In a general equilibrium model all prices are variable and all markets clear. The relationships and feedback within and between markets are only considered in a general equilibrium model. Both partial and general equilibrium models may be used to analyse environmental issues.

Externalities that are caused by production or consumption, such as pollution, may affect utility, production and social welfare. In an equilibrium the (negative) externalities may be optimized by imposing policies. With these policies the market equilibrium equals the social welfare optimum. Chapter 7 gives a theoretical equilibrium model in an M-P framework. In this general equilibrium model the market equilibrium, that is obtained by optimizing the behaviour of various producers and consumers, does not include the externalities of using new materials and of pollution. To achieve a situation where the market equilibrium equals the social welfare optimum, taxes need to be incorporated in the market equilibrium. Besides theoretical models there is a growing number of applied equilibrium models.

Important questions which are considered in the vast literature on externalities are very varied and may be arbitrarily listed as follows.

- Which policy instruments may be imposed in which situation?
- On the basis of which criteria should environmental policy instruments be selected (see Section 3.5.2)?
- At what scale level (e.g. national or international) should environmental policy be implemented (Baumol and Oates, 1988)?
- What are the uncertainties regarding pollution (Lines 1995)? For example, what are the possible effects of pollution on the environment?
- How should pollution in an imperfect market, such as an oligopoly or a monopoly be considered (Carraro *et al.*, 1996)? In the situation of an imperfect market the optimization of an externality may have effects on the level of production or output.
- How can imperfect knowledge or information to be handled, e.g. about the pollution caused by each agent (Carraro *et al.*, 1996)?

Pollution models and material flows

The few pollution models which include material flows consider these flows in monetary terms instead of physical terms (Wertz, 1976; Sullivan, 1987; Copeland, 1991). Just a few studies include the physical or environmental aspects of these flows (Dinan, 1993; Keeler and Renkow, 1994). Pollution models normally incorporate direct and indirect substitution in the production function. Recycling may be included in pollution models (Fullerton and Kinnaman, 1995).

The material balance principle is normally not included. For example, the production function $Q=f(k,r,v)$ with inputs capital (k), recycled and new materials (r and v) does not say anything about the material contents of the product Q. For the transformation of the product Q into waste material or materials to be recycled it is necessary to keep track of these contents. This is done in Chapters 6 to 9 in Part III of this study.

Here, some examples are given of optimization and equilibrium models in which material flows are explicitly considered. Keeler and Renkow (1994) minimize the costs of solid waste management under a set of physical constraints of recycling and waste generation. These physical constraints are directly related to the costs of recycling and costs of incineration. The only material balance condition included is that all the waste generated needs to be recycled, incinerated or landfilled. Dinan (1993) analyses policies for (paper) use and waste, maximizing social welfare. New and recycled resources are incorporated in the production function, recycling and disposal costs, and all the material balance conditions of the material flows are included in the model. It is concluded that a tax on new materials is not efficient, but a combined disposal tax and a reuse subsidy is.

The connection between resource use and the disposal of waste, and recycling has been studied by Fullerton and Kinnaman (1995) in a general equilibrium model. Here, various scenarios are considered in which the negative impacts of recycling or the production of a substitute are taken into account. To analyse various environmental policies dealing with solid waste generation of households, the utility of the consumers is optimized subject to a household production function, a time constraint (leisure or production activities) and a budget constraint under various environmental policies for solid waste generation (Morris and Holthausen, 1994).

Deacon (1995) analyses the relationship, in a static general equilibrium framework, between government policy and deforestation, comparing a social welfare optimum and a market equilibrium to obtain optimal policy rules. The forest itself directly provides a utility and indirectly by the goods which are produced by using the wood of the forest. The decision to use or preserve the natural resources (the forest) is clearer if the stock of resources is included in the utility function. An example of a partial dynamic equilibrium model is that developed by Deacon (1993). It simulates the effects of taxes on petroleum exploration and production. Taxation may cause low grade resources not to be extracted and it may change the allocation of production over time. The impact of a subsidy on the inputs of natural resources in a general equilibrium model with two sectors and three production factors is analysed in Hertel (1988).

An example of a general equilibrium model

A simple static general equilibrium model is presented to illustrate material flows in a closed economy with 2 producers, 2 types of goods and a consumer (partly based on Dinwiddy and Teal, 1988, Chapter 2). The material balance principle for resources and waste generation is included in the model. In the general equilibrium model, the consumer maximizes his utility, U, which depends on the number of the two types of goods, C_1 and C_2, and the amount of waste generated, W. The effect of W on the utility function of the consumer is the externality. The consumer is to determine his choice on:

$$\max U = U(C_1,C_2,W)$$

with negative marginal utility of waste (W) and positive marginal utilities of the consumption goods (C_1 and C_2). Waste (W) may be interpreted as a stock or a flow, because in a static model there is no distinction. The utility is maximized under the budget constraint with price p_i for good C_i, and the total income Y:

$$\sum_{i=1}^{2} p_i C_i = Y$$

From this optimization problem demand functions can be derived which depend on the prices of the good, the total income and the waste generated.

$$C_i = C_i\ (p_1, p_2, Y, W) \text{ for } i=1,2$$

The producers maximize their profit, Π_i, which equals the revenue of the products that are sold, $p_i X_i$, minus the costs of the inputs (fixed and given p_k and p_m), given their production function f^i with inputs capital (K_i) and materials (M_i):

$$\max \Pi_i = p_i X_i - (p_k K_i + p_m M_i) \text{ for } i=1,2$$

$$X_i = f^i\ (K_i, M_i) \text{ for } i=1,2$$

The supply functions for goods can be derived from this optimization:

$$X_i = X_i(p_i, p_k, p_m) \text{ for } i=1,2$$

The demand functions for capital and materials are:

$$K_i = K_i\ (X_i, p_k, p_m) \text{ for } i=1,2$$

and the demand for materials:

$$M_i = M_i\ (X_i, p_k, p_m) \text{ for } i=1,2$$

The commodity or goods market is cleared when the demand for goods, C_i, equals the supply of goods, X_i for $i=1,2$. Thus, $C_i=X_i$ for $i=1,2$. The factor market for capital is cleared when the amount of capital for both producers equals the total (fixed) amount of capital (K^*), $K_1+K_2=K^*$.

The income formation of the consumers is the sum of the production factors capital and materials, and profit.

$$Y = \sum_{i=1}^{2} p_k K_i + p_m M_i + \Pi_i$$

The total amount of materials used is $M=M_1+M_2$. The amount of materials used (M) equals the amount of waste, W. This is the material balance condition $M=W$, that is the final equation of this model, 'closing' the materials side of it.

This model presents a neoclassical general equilibrium model that considers the material balance conditions and material flows explicitly. This example shows that material flows can be integrated in a neoclassical economic model.

Applied general equilibrium models

Applied or computable general equilibrium (AGE/CGE) models can measure the economy-wide effects of policies on production sectors and household groups, or more specifically on production sectors and household groups. AGE models are complex and require a large amount of data. In AGE models the quantities and the relative prices of product groups are determined within the models. This differs essentially from the optimization (or planning) models (see Section 5.2) and from the input-output models which will be described in Section 5.4. AGE models aim to solve the equilibrium of the allocation patterns of production and consumption (Bergman, 1995).

AGE models are very diverse in their geographical scale (global, national), their level of aggregation (geographical zones, households, products), their time scale (static, dynamic), and their inclusion of environmental variables (emissions, energy, materials) (Destais, 1996). Most AGE models that deal with environmental issues are comparative static and consider one or more types of emissions on a national level.

Single country AGE models primarily deal with national policies. A static AGE model has been applied to Sweden to evaluate the impact of policies aimed at reducing sulphur emissions in Sweden on productivity and welfare (Bergman, 1995). A static AGE model for the Netherlands has been developed to analyse the socio-economic impacts of an energy tax on households or firms (Dellink and Jansen, 1995). In a comparative static setting various policy scenarios for CO_2 emissions have been analysed and compared. Dynamic and global models are scarce. An example of such a global and dynamic AGE model is GREEN. This model can evaluate the economic costs of international agreements to reduce global emissions of carbon dioxide, CO_2 (OECD, 1992). For a survey of AGE models see, for instance, Bergman *et al.* (1990).

These AGE models mainly deal with the emissions of energy or toxic materials. Material or energy flows are mainly modelled in the production and consumption stages without dealing with extraction. Emissions are a result of the production and consumption processes.

A special topic in AGE models is whether or not environmental taxation leads to a 'double dividend', i.e. positive effects on both environment and employment (for theoretical views and models, see Bovenberg and De Mooij, 1994; Bohm, 1997). A general conclusion is that a double dividend depends on the assumptions about mobility of labour and capital. Several AGE models have been applied to study the possibility of a double dividend when imposing an energy tax (OECD, 1992; Conrad and Schröder, 1993; Jorgenson and Wilcoxen, 1993; European Commission, 1993; Dellink and Jansen, 1995). AGE models may also be used to study the effect of a material tax and the issue of a double dividend. In Chapter 10 of this study an AGE for the Netherlands will be applied to study the effects of material policies.

Aspects of pollution models
Table 5.2. gives an overview of the dominant aspects of pollution models. These characteristics should be seen as indications of the type of model. Chapter 6 will give an application of an optimization model in which the costs are minimized. Chapter 7 presents a general equilibrium model with material flows.

Table 5.2. The characteristics of pollution models.

Goal of the model	Optimal regulation of polluting economic activities
Type of model	Optimization (including equilibrium models)
Time aspect	Static and occasionally dynamic (equilibrium models) Static and dynamic (optimization models)
Spatial scale	Dependent on environmental problem, but mainly on a national level
Aggregation level	Sectoral or national level
Units	Prices and utility

5.4. Environmental Input-Output Models

Further to the description of input-output (I-O) models with resources and pollution in Section 4.4 of Chapter 4, this section focuses on environmental I-O models.[3] Augmented and integrated environmental-economic I-O tables are discussed. This section is longer than the one in Chapter 4 because I-O modelling is relevant for many environmental questions and has a connection with material flow models (see Chapter 4).

The environmental sectors, such as the abatement sectors, are added to an I-O table which makes it an environmental extension of the economic I-O table, known as an augmented or extended I-O table (Leontief, 1970; Victor, 1972; Leontief and Ford, 1972). An augmented I-O table can include environmental inputs, such as minerals or forestry, discharges to the environment, etc. (see Table 4.1). Furthermore, an abatement sector may be included as shown in Table 5.3.

The abatement sectors themselves also generate pollution (A_4Z) and the provide inputs for the production sectors (A_3X). The input of the abatement sectors is the pollution generated by the production sectors (A_2Z) and a part of the pollution generated by the abatement sectors themselves (A_4Z). The term Z is the vector of the output of the abatement sectors and A_1X is the matrix of the input coefficients (A) times the vector of the output of the production sectors (X). The pollution abatement coefficients reflect inputs to pollution abatement activities.

[3] Early applications of I-O analysis to environmental policy are Ayres and Kneese (1969), Kneese *et al.* (1970), Victor (1972), James *et al.* (1978). More recent static applications are: Huang *et al.* (1993) on solid waste, Konijn (1995) on primary materials and energy carriers, and Weber (1995) on energy requirements in Germany.

Table 5.3. An augmented I-O table.

Input / Output	Production sectors	Abatement sectors	Final demand	Total input	Total output
Production sectors	A_1X	A_2Z	F_1	--	X
Abatement sectors	A_3X	A_4Z	$-F_2$	--	Z
Primary inputs	M_1X	M_2Z	--	E	--

It is important to distinguish between an I-O table or accounting framework and an I-O model. A table or an accounting framework is a description of the physical or monetary flows between different sectors of the economy. It provides insight into the interrelations of these sectors. For an I-O model the accounting framework is the base for the analysis of various changes in the inputs, outputs or coefficients. An I-O model may be part of an optimization model in which, for instance, the emissions are minimized or the value added is maximized.

The augmented I-O table (Table 5.3) may be translated into the following model:

$$A_1X + A_2Z + F_1 = X$$
$$A_3X + A_4Z - F_2 = Z$$
$$M_1X + M_2Z = E$$

The vectors of the output level of the production sectors (X) and the abatement sectors (Z), and the matrices of the fixed technical coefficients (A_1-A_4 and M_1-M_2) determine the final demand of the production sectors (F_1) and the final demand of the abatement sector ($-F_2$), to be interpreted as the socially (tolerated) pollution level. Or the other way around, the demand of the production sectors (F_1) and the pollution level ($-F_2$) determine X, Z, F_2 and E.

The total amount of pollution eliminated (abated) is Z. The matrices A_3X and A_4Z equal the pollution generated by the production and abatement sectors. The required primary inputs are the sum (E) of the required inputs in the production and abatement sectors (M_1X and M_2Z). Rewriting the first two equations gives:

$$\begin{pmatrix} X \\ Z \end{pmatrix} = \begin{pmatrix} 1-A_1 & -A_2 \\ -A_3 & 1-A_4 \end{pmatrix}^{-1} \begin{pmatrix} F_1 \\ -F_2 \end{pmatrix}$$

This augmented I-O table can be used to analyse various interactions between the use of natural resources, the output of the production and abatement sectors and the permitted or resulting pollution. For example, with a given level of permitted pollution, $-F_2$, and a given final demand, F_1, the required output, X, and abatement level, Z, can be calculated. Then the primary inputs, E, needed for X and Z can be calculated using the third equation, $M_1X+M_2Z=E$. A row of environmental inputs may be added to Table 5.3. Another example of this extended I-O model is for the calculation of the required natural resources (E) and the total level of pollution ($-F_2$), given total output levels of both types of sectors (X and Z).

An integrated environmental-economic I-O accounting framework
Table 5.4 shows the general structure of an integrated environmental-economic I-O table. It includes the following flows: resources from the environment to the economic sectors (G); interactions between economic sectors (A); emissions from the economy to the environment (B); and, interactions between environmental sectors (H) (Cumberland, 1966; Daly, 1968; Isard *et al.*, 1971).

Table 5.4. An integrated economic-environmental I-O table.

	Economic sectors	Environmental processes
Economic sectors	A	B
Environmental processes	G	H

These integrated environmental economic I-O tables are very difficult to implement mainly because of their large data requirements. The interactions between the environmental processes (H) is often not included in the table.

Dynamic I-O modelling
To convert a static I-O table in a forecasting model, a future path of final demand needs to be fixed (Ayres, 1978). Some assume that I-O tables can be used for a period of 5 years (James *et al.*, 1978) and others for 10-15 years (Fankhauser and McCoy, 1995). When I-O tables of various years are available they may provide information on the changes in monetary or physical terms and they may be used for forecasting or scenario-analysis. Some examples of applications of dynamic I-O tables or models concern structural changes and energy consumption in Japan (Han and Lakshmanan, 1994), the impact of definite policies using time-series analysis (Midmore, 1993), and supporting growth policy (Henry, 1994).

Although there are problems of data availability and measurement, dynamic I-O models have been used to analyse the impacts of pollution-control technologies on a regional scale (Miernyk and Sears, 1974). Duchin and Szyld (1985) use a dynamic I-O model to analyse the impacts of computers and automation in the US. Hamilton (1997) uses basically the same dynamic I-O model as Duchin and Szyld (1985) and applies it to Indonesia. The investments in the model are endogenous depending on expected growth and production technologies. One of the results of the analysis is that the logging industry in Indonesia is not sustainable, albeit that this is often claimed. For more discussion on dynamic I-O analysis, see James *et al.* (1978), Miller and Blair (1985), Duchin (1988) and Idenburg (1993).

Applications of augmented I-O models to the environment
The augmented I-O table is difficult to apply. Leontief and Ford (1972) calculated the emissions of five air pollutants per industrial sector in the United States. These were calculated in physical units related to monetary output units. Forsund (1985) used an extended I-O table to calculate 37 types of emissions for the Norwegian economy. McNicoll and Blackmore (1993) calculated the emissions for 12 pollutants in Scotland. For the Netherlands an augmented I-O table with energy flows and

pollutants was constructed. Hafkamp (1991) gave a short overview of this I-O model (called RIM). Dellink *et al.* (1996) combined an augmented I-O table and linear programming for the development of various future scenarios for the Netherlands. Here, an I-O table was extended with an environmental module with natural resources (matrix G in Table 5.4) and emissions (matrix B in Table 5.4). In this I-O model matrix H of Table 5.4 was not included. In the I-O table ten types of physical emissions were included. Luptácik and Böhm (1994) used the augmented I-O table for theoretical optimization models with multi-criteria analysis. The weights may be interpreted as emission charges.

Various countries include environmental inputs and outputs in their national accounts. This topic is beyond the scope of this study (for an overview on the use of various systems of environmental national accounts, see Lintott 1996). In the Netherlands the National Accounting Matrix including Environmental Accounts (NAMEA) is an accounting framework in which environment and economy are combined (Hueting, 1991; De Boo *et al.*, 1991; De Haan and Keuning, 1994). Pearson (1989) discusses the use of augmented I-O tables for proactive policy strategies which focus on the anticipation and prevention of environmental problems, instead of the reactive policy strategies that deal with solving existing problems.

Duchin and Lange (1994) performed a dynamic I-O study based on the World Model of Leontief, Carter and Petri (Leontief *et al.*, 1977). Besides monetary flows (for example, capital flows and trade of commodities) this study includes the use of materials (for example, metals, cement, paper, chemical) and energy, and the emissions of several pollutants (CO_2, SO_x and NO_x). The results of a reference scenario with no technological change are compared with those of an 'Our Common Future' (OCF) scenario (based on WCED, 1987) that includes technological development. It is concluded that under the OCF scenario global consumption is higher than under the reference scenario, but the distribution of consumption is more uneven in the OCF scenario. Under the OCF scenario emissions are lower than under the reference scenario, but the emissions still increase.

An example of an augmented I-O table and analysis

A numerical example of an augmented I-O table is given to illustrate the possible analyses of economic and environmental sectors and their interactions using these I-O tables. In this example of an augmented I-O table, the economy consists of four sectors. The two economic sectors are food production and production of machines, and the two environmental sectors are highly polluted water and slightly polluted water (adapted from Duchin, 1992). The basic equations of a physical and monetary I-O table are: (I-A)x=y, (I-A')p=v, p'y=v'x, with the matrix of I-O coefficients (A), the output (x), the final demand (y), the prices (p), and the value added per output (v).[4] The matrix of I-O coefficients is:

[4] Note that in this example the physical variables x (the output) and y (the final demand) are converted into monetary variables by multiplying with the value added per output (v) and the prices (p). The multiplication of the vectors v and x, v'x, and the multiplication of the vectors p and y, p'y, equal the total income.

$$A = \begin{bmatrix} 0.4 & 0.3 & 0 & 0 \\ 0.2 & 0.3 & 0.2 & 0.1 \\ 0.4 & 0 & 0 & 0 \\ 0.4 & 0.1 & 0.3 & 0.2 \end{bmatrix}$$

The final demand is: y'=[10 3 0 0], and the value added is: v'=[10 10 50 20]. The final demand of the highly and slightly polluted water is zero (see last two numbers of y'), which means that all waste water is treated. With these above data, the total output, the unit prices and the total income are calculated: x'=[25.2 17.0 10.1 18.5], p'=[110 66 73 33] and p'y=v'x=1294. The residuals, s, are assumed to be: s'=[0.05 0 0.05 0.02]. Therefore, the total residual are s'x=2.1.

In a scenario in which the final demand is: y'=[10 3 -2 -2], the level of allowed dumping of highly polluted and slightly polluted waste water equals 2. Assume A, s, v to remain equal. The total output, the unit prices, the total income and the total residuals then become: x'= [24.5 15.7 7.9 14.6], p'=[92 58 70 31], p'y=v'x=1092, s'x=1.9. In this numerical example the effect of less treatment of polluted water is: (i) lower prices and lower costs (p' and p'y); (ii) less other residuals (s'x); and, (iii) more polluted water (from 0 to 2).

Environmental I-O analysis and material flows

An interesting question concerning material flows that may be answered with I-O models is: How much materials are used or waste generated directly and indirectly by a product, a sector or a final demand category? (Ayres, 1978). Most I-O tables are in monetary units and to obtain the physical flow of materials these monetary units need to be converted, which is complicated because of the usual aggregated index of goods. When environmental aspects are included in an I-O model, the same problem holds. For a physical I-O model the use of materials may be derived straightforwardly, but to obtain such an I-O model the transformation from monetary to physical units is needed (see, for an example of a transformation model, Wieringa and Van den Nieuwenhuijzen, 1994). In Section 4.4 of this study physical I-O models are discussed. Some argue that in a monetary I-O table the rows and columns may be added because all are in monetary terms, while in a physical I-O table the different inputs going into each sector cannot be added (Leontief, 1986; Heijungs, 1997). However, it may be argued that if all physical units are measured in kilograms, the rows and columns may be added and the totals balance.

Important aspects of material flows that are generally not included in monetary I-O models are technological change, substitution, material balances and recycling. Technological change, including for example the use of other, or less, materials may only be incorporated in dynamic I-O models, because in static I-O models the technological coefficients are fixed. Substitution in I-O modelling is impossible, once again because of the fixed coefficients. The material balance approach which has been promoted by Kneese *et al.* (1970) and Ayres (1978) is rarely followed (see also discussion in James, 1985; Pearson, 1989). The introduction of a recycling sector in an I-O table is possible, but the technological coefficients in an I-O table are fixed and therefore the production of the goods requires fixed amounts of new and recycled inputs. Thus, no substitution between new and recycled inputs exists. Therefore, the inputs of the production, recycling and abatement sectors are partly from new (primary inputs) and partly from recycled (recycling sector) inputs.

The application of I-O modelling to material flows is also helpful in tracing back these flows (Ayres, 1978; Duchin, 1992). The trade-off between environmental effects, physically or in externality terms, of new and recycled inputs is important. Duchin (1992) discusses the use of I-O modelling for 'industrial ecology' focusing on strategies to reduce and recycle materials over time. The questions which need to be solved are: How can the reduction and recycling be reached? How much will be the costs and benefits of reduction and recycling? And, which policies are needed? To answer these questions needs a simultaneous physical and economic analysis for which environmental I-O analysis may be used (Duchin, 1992).

An augmented I-O model with recycling and substitution

In standard I-O models the choice between using new or recycled inputs is not included. For material flow analysis this choice is very important for the environmental effects of resource use and pollution. A combination of an I-O model with an optimization principle based on monetary or physical units allows the effects of different combinations of, for example, new and recycled inputs, to be analysed. In a physical I-O model with recycling, the material balance approach may be followed (Kneese *et al.*, 1970; Ayres, 1978; James, 1985; Duchin, 1992).

Table 5.5 presents an example of an I-O table with recycling and substitution between new and recycled materials. In order to make substitution possible between primary inputs, i.e. new materials, and the inputs of the recycling sector, variables α and β are introduced. These variables may be interpreted as the parts of the output made by new (β) or recycled (α) materials. One unit of output X needs a combination of new and recycled materials: αB_1 and βM_1. Changes in the variables α and β may be interpreted as substitution between new and recycled materials. Thus, if α decreases and β increases then less new materials but more recycled materials are used by the production sector, implying substitution between those two types of materials. If β directly depends on α, say $\beta = 1-\alpha$, then a decrease in the use of new materials (α) results in an increase in the use of recycled materials (β). In an optimization framework α and β are choice variables.

Table 5.5. Augmented I-O table with recycling.

	Production sector	Recycling sector	Abatement sector	Final demand	Total output (Total input)
Production sector	A_1X	A_2Q	A_3Z	F_1	X
Recycling sector	αB_1X	B_2Q	B_3Z		Q
Abatement sector	C_1X	C_2Q	C_3Z	$-F_2$	Z
Primary inputs (new materials)	βM_1X	M_2Q	M_3Z		(E)

The I-O model based on Table 5.5 may be written as:

$$\begin{bmatrix} X \\ Q \\ Z \end{bmatrix} = \begin{bmatrix} 1-A_1 & -A_2 & -A_3 \\ -\alpha B_1 & 1-B_2 & -B_3 \\ -C_1 & -C_2 & 1-C_3 \end{bmatrix}^{-1} \begin{bmatrix} F_1 \\ 0 \\ -F_2 \end{bmatrix}$$

In addition, together with the equation for the primary inputs: $\beta M_1X+M_2Q+M_3Z=E$. Assuming that the primary inputs and the recycling sector outputs are the only material inputs, the equation of new and recycled material inputs then becomes (with X^* the output of the production sector in physical units): $\alpha B_1X^*+\beta M_1X^*=X^*$. Given the final demand, F_1, and the allowed level of pollution, F_2, feasible combinations of the outputs X, Q, Z and E and the choice variables α and β may be calculated according to an optimization principle, such as the minimal costs of generating F_1 and F_2, or maximum recycling. These choice variables may be restricted or related to each other: for example, by a technological restriction such as $2\beta<\alpha$. This restriction indicates that the use of recycled materials is always less than half of the use of new materials, implying that for the production at least one-third of the materials need to be new materials.

This I-O table may be in physical, monetary, or in physical and monetary terms with the same matrix of technological coefficients (here A_i, B_i, C_i and M_i for $i=1,2,3$). The transactions between economic sectors may be either in monetary or in physical terms. The generation of pollution, for example, is in physical terms. Thus, the optimization model may be in physical and monetary units. An example is to minimize the total costs of meeting the demand (in monetary terms) under the constraint that the total pollution (in physical units) may not exceed a certain amount.

A commodity-by-industry table[5]

I-O tables are based on data in which every firm is assigned to an industrial category (the SIC-code in the US and the SBI-code in the Netherlands) according to its primary product. These conventional I-O tables are industry-by-industry tables. This means that it is assumed that every economic sector produces one good, while in reality many economic sectors do not produce only one good but several (multi-output of products), which makes the assignment of the whole output to one good or commodity inaccurate. Therefore, an alternative is to use commodity-by-industry (C-I) tables which consist of a 'use table' and a 'make table' in monetary terms. C-I tables are normally in monetary terms to allow for addition over industries. The C-I tables are more accurate when economic sectors produce several products (or commodities). If each economic sector produces one single product, C-I tables reduce to I-O tables.

[5] A commodity-by-industry table may also be referred to as a make-use table.

Economic-environmental commodity-by-industry table
The C-I table can be augmented with environmental inputs according to the approach of Victor (1972) and by pollution and abatement (Idenburg and Steenge, 1991). Table 5.6 shows the commodity-by-industry approach with the environmental commodities as added by Victor and with pollution and abatement. In comparison with the integrated economic-environmental I-O table (see Table 5.4) the economic-environmental C-I table is easier to implement because the data needed are available directly from the industries, i.e. no transformation is needed to assign every sector to one good (Miller and Blair, 1985). Victor (1972) introduced the C-I table as presented in Table 5.6.

Economic subsystem:
U = use matrix: the amount of economic commodities the industries use
V = make matrix: the amount of economic commodities the industries make
F = vector of final demand for economic commodities
Q = vector of the gross (i.e. total) output of economic commodities
W = vector of the value added inputs of the industries
X = the vector of total output of industries

Environmental subsystem:
R = matrix of environmental commodities discharged as a result of the production of the commodities
S = matrix of environmental commodities discharged by the industries
P = matrix of environmental commodities used by the production of commodities
T = matrix of environmental commodities used by the industries

With this C-I table the environmental and economic input requirements can be analysed for the various industry and commodity levels. The advantage of using a C-I table is that the linkage between the production of commodities and the use and discharge of environmental inputs is more accurate than in an I-O table.

Table 5.6. Commodity-by-industry table with environmental inputs, pollution and abatement (adapted from Miller and Blair, 1985).

	Commodities	Industries	Final demand	Total output	Environmental commodities
Commodities		U	F	Q	R
Industries	V			X	S
Value added		W	GNP		
Total inputs	Q^T	X^T			
Environmental commodities	P	T			

Activity analysis

A method which is strongly related to I-O modelling is activity analysis. A multi-activity static optimizing model (in monetary terms) was invented simultaneously by Kantorovich and Koopmans in 1951 to eliminate a drawback of I-O analysis, namely the fixed technological coefficients. In I-O analysis a commodity (good or service) is only produced by one sector and the final demand of a sector equals the total output minus the intermediate consumption. In an activity analysis, however, several activities or processes may produce one commodity, while also one activity may produce multiple commodities. Also in C-I tables the activities may produce one or more products, but in C-I tables the coefficients are fixed and in activity analysis these are chosen. This allows the optimal allocation of production capacities to be analysed. Each activity needs a certain level of inputs per unit of output. Activity analysis usually measures commodities in monetary terms, but they can be expressed in physical terms too. This allows the physical relationships between commodities, stocks and flows of materials and the physical laws to be taken into account.

An activity analysis production function may be seen as a combination of a linear and a Leontief (fixed input) production function. Figure 5.1 presents three types of production functions with two inputs: a Leontief, a linear and an activity analysis production function (based on Ayres, 1978).

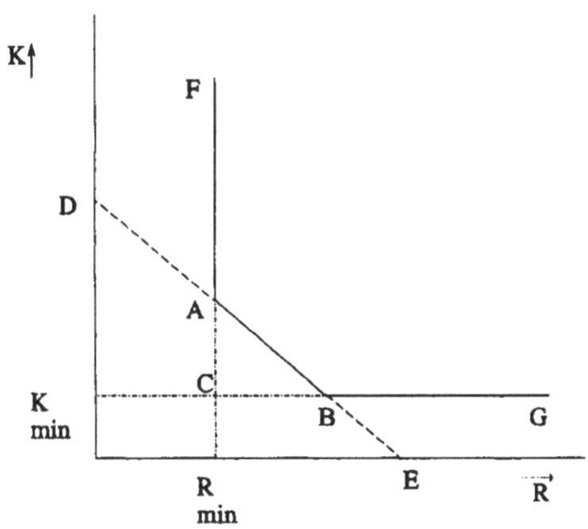

Figure 5.1. An activity analysis production function.

The Leontief production function (F-A-C-B-G) shows that there is no substitution possible between the two inputs. This is the type of production function that is standard for I-O analysis (see Section 5.4). The linear production function (D-A-B-E) shows that substitution is unlimited. Without the input of R, production is still possible (point D). The activity analysis production function (F-A-B-G) shows that a certain minimal amount of inputs is needed for one output (K min and R min in Figure 5.1). This implies that (linear) substitution is only possible with at least the minimal level of every input. Thus, only between point A and B is substitution possible. The activity analysis function is piecewise linear and concave.

Originally, the goal of activity analysis was static optimization of the amount of inputs or waste, or the allocation of energy using the choice between different technologies (Ayres, 1978). More recently, the idea of a limited amount of inputs needed to produce a certain good has been adapted to environmental inputs. Instead of requiring a minimal level of input, a maximum level of inputs, or a maximum level of (undesirable) outputs, is set (Färe *et al.*, 1996). Since the middle of the 1970s activity analysis has no longer been used extensively.[6]

Aspects of environmental input-output tables.
A short overview of some characteristics of environmental I-O tables is given to compare these type of models with other models dealt with in this chapter.

Table 5.7. Characteristics of environmental I-O models.

Goal of the model	Description of the sectoral structure of the economy
Type of model	Descriptive (or embedded in an optimization framework)
Time aspect	Static (sometimes dynamic)
Spatial scale	National or regional
Aggregation level	Sectoral
Units	Monetary (value) and physical terms

5.5. Macroeconomic Models

Macroeconomic models are mainly used to forecast economic performance and to analyse the impact of policies on the (national) economy. They consist of a set of equations and identities. Macroeconomic models may provide predictions of macroeconomic factors, such as GNP, unemployment, inflation, balance of payments and growth. Macroeconomic models can also be used for the analysis of environmental policies which have an economy-wide impact, for example energy policies. For sector or product specific policies which do not substantially influence the national economy, these models are less appropriate.

Macroeconomic models do not explicitly maximize welfare, which is a main difference with general equilibrium and welfare-based optimization models, as discussed in Section 5.3. Macroeconomic models may be estimated by empirical data over time which may make them less abstract than, for example, applied general

[6] A special type of model which is a mixture of activity analysis and general equilibrium (GE) models is the 'activity analysis general equilibrium model' (Bergman, 1990). This model may be classified as a variant of GE, although based on a linear programming format. Activity analysis GE models have been applied to energy in Manne (1977) and to international trade by Ginsburg and Waelbroeck (1981 and 1984).

equilibrium models, which are usually based on static benchmark data sets. Furthermore, macroeconomic models do not assume equilibrium and market clearing, as general equilibrium models do. In macroeconomic models a temporary, or even a persistent disequilibrium may therefore exist.

Macroeconomic models and material flows

Macroeconomic models may be used to describe the relationship between growth and environmental pressure. Macroeconomic models may be used to analyse the (sectoral) effects of investments in the environment or environmental requirements on production, investment and employment (Den Hartog and Maas, 1990). They may also be used for forecasting and when policy analysis of environmental issues is performed. For example, in several countries in the European Union the macroeconomic effects of energy taxes have been studied (European Commission, 1992). Dynamic or intertemporal issues may be dealt with in macroeconomic models, for example to predict the interest rate (Fankhauser and McCoy, 1995).

Macroeconomic models are not very appropriate for studying material flows because of their high level of aggregation and the omission of physical variables. However, macroeconomic models may generate (macro-) scenarios which may serve as an input for other models.

Environmental Kuznets Curve

The connection between economic growth and environmental pressure may be described with an environmental or green Kuznets curve (EKC). Some of the literature claims that economic growth (income) and environmental pressure are positively correlated until income reaches a certain level beyond which the relation is negative, a situation known as 'delinking' (Selden and Song, 1994; Common, 1995). Then, the relation between income and environment has an inverted U-shape. However, De Bruyn and Opschoor (1997) found indications that after a period of delinking, a new period of 'relinking' may be shown. Heintz and Verbruggen (1997) give an overview of these studies and of the conditions under which the described correlation is valid. Ekins (1997) examines various studies on the EKC and concludes that the inverted U-shape relationship is only found in some studies and for some environmental indicators. There are no econometric studies of the causation of the EKC relationship. Therefore, it is suggested that the negative relationship between environmental quality and income is more likely to result from environmental policy than from endogenous changes in economic structure or technology (Ekins, 1997).

Economic growth and the environment

Neoclassical growth models are a special group of macroeconomic models based on neoclassical theory. Growth models may be classified as either exogenous or endogenous (for overviews of growth models, see Toman *et al.*, 1995; Barro and Sala-i-Martin, 1995). In this section, only exogenous growth models will be discussed very briefly. Endogenous growth models in which technological change is important will be examined in Section 5.6.

In the Solow-Swan model the production function is $Y(t)=F(K(t),L(t),t)$ with production (Y), capital (K), labour (L) and time (t). Investment depends on the savings rate of households (s), the rate of technological progress (δ) and the population growth (n). The main equation of this model is:

$$\dot{k} = s\ f(k) - (n+\delta)k$$

with

$$\dot{k}=\dot{K}/L-nk\ ,\ n=\dot{L}/L$$

In exogenous growth models the long-term steady-state growth rate (which is equal for all variables) is defined as the point where k is constant. The long-run growth rates are exogenous in this type of model. However, the transition towards this long-run growth rate is interesting.

In a Solow-Swan model with exogenous growth with natural resources the production function is: Y=F(K,N) with production (Y), capital (K), natural resources (N). The use of natural resources depend on the labour supply (L), the exogenous labour efficiency, π, which is increasing in time t, $N=\exp(\pi t^*L)$. The optimal growth path with exhaustible resources is examined (Solow, 1974 and 1986).

In an exogenous growth model with pollution, the latter is seen as an unavoidable by-product of consumption which may be reduced by abatement activities. Therefore, society has to choose between consumption, growth and abatement (Gradus and Smulders, 1993). Exogenous growth models are mainly analytical (optimization) models. The natural resources and pollution are mainly considered in externality (or monetary) terms. Physical variables may be included, for example the effect of pollution on production or the production function. For studying material flows physical variables and thermodynamic constraints, including material balance conditions, need to be added to these models (see the discussion on production function and material flows in Section 5.2).

Aspects of macroeconomic models

To compare macroeconomic models with other types of models, some characteristics of macroeconomic models are listed in Table 5.8.

Table 5.8. The characteristics of macroeconomic models.

Goal of the model	Forecast economic performance and impact of policies
Type of model	Descriptive (including forecasting)
Time aspect	Dynamic (5-7 years forecasting)
Spatial scale	National
Aggregation level	Sectoral
Units, variables	GNP, growth rate, unemployment rate, inflation rate, balance of payments

5.6. Models of Technological Change and Economic Evolution

This section deals with models in which technological change may be dealt with explicitly. First, a concise view of the concept of evolutionary economics is given. Evolutionary economics uses evolutionary and developmental concepts which are different from the concepts of optimization and equilibrium mainly used in the other modelling approaches described in this chapter (Sections 5.2 to 5.5). Furthermore, possible applications of evolutionary economic model types to material flows are briefly discussed. Second, endogenous growth models are discussed. Although these models are mainly dynamic neoclassical equilibrium models, they are included in this section because technological change is their main feature.

In Section 3.4.4 the distinction between gradual and radical technological changes was made. This gradual change may also be called predictable (phenotypic) and unpredictable (genotypic) change. In the equilibrium concept of neoclassical economics technological change is seen as a change from one equilibrium to another. In reality though, technological change spreads slowly making an evolutionary concept more accurate (Nelson, 1995). In mechanistic models (see Sections 5.2 to 5.5, and also endogenous growth models, discussed later in this section) only predictable technological change may be taken into account. In non-mechanistic models, such as evolutionary models, unpredictable or undirectional technological change may also be considered explicitly (Dosi *et al.*, 1988; Freeman and Perez, 1988; Kemp, 1995). The processes of diffusion and adoption are also studied (Kemp, 1995).

Evolutionary economics is based on the idea that the economy is a part of human activity and society which is (structurally) changing over time. Production and consumption of economic goods may be seen as part of the global ecosystem or in other words, 'the economic system is a subset of the biophysical world' (Gowdy, 1994). Ecological interaction provides a selective mechanism which may result in the disappearance of goods for which there is no demand and in the invention of new goods (Boulding, 1981). Many economic phenomena appear to have a counterpart in biology and therefore an evolutionary perspective may possibly be applied to economics as well. An evolutionary perspective may open the way towards a non-mechanical economics. This is in contrast with the mechanistic, neoclassical I-O and macroeconomic models. Because the main focus of evolutionary economics is on changes over time, the concept of equilibrium that is often used in economics is not appropriate for describing or analysing changes over time, because economies may be in a disequilibrium.

The breakthrough of evolutionary thinking in economics was the book by Nelson and Winter (1982). They were strongly influenced by the ideas of Schumpeter who saw 'capitalism as an engine of progressive change' (Nelson and Winter, 1982, p. 39). Competition and innovation are basic elements for Schumpeter (1934) and the evolutionary thinking of Nelson and Winter (1982). Nelson and Winter systematically describe the need for an evolutionary approach in economics because neoclassical economic theory that is based on optimization and equilibria does not adequately consider learning processes, self-adaptation, self-organization, path-dependency, anticipation of future developments (Nelson, 1995). With an evolutionary approach non-deterministic, non-mechanistic and unpredictable processes or changes may be considered (Nelson and Winter, 1982).

Allen (1996a and 1996b) states that - as in neoclassical economics - dynamic models may result in predictions or simulations over time, but that these deterministic or mechanistic models are not capable of including qualitative changes. This implies that a deterministic model is only correct as long as the system does not change qualitatively. Several models have been developed which include non-deterministic processes and non-average behaviour of agents. For long-term processes an evolutionary approach seems more appropriate.

In neoclassical theory, real ecological problems are ignored: especially the irreversibility of many natural phenomena cannot be taken into account (O'Connor, 1993). This irreversibility is another important issue in evolutionary economics. Irreversibility exists on two physical levels: the current and the future use of energy and materials in which the future needs to be seen in the light of inventions and technological change. First, the production process has a physical base to which the laws of thermodynamics apply (see Chapter 2). Here, the material balance approach and the second law of thermodynamics, the problems of extraction and pollution and the limited possibilities of recycling need to be taken into account. Second, an invention is an unpredictable, but irreversible process. An invention, either a resource-saving or a resource-substituting invention, is irreversible in time. O'Connor (1993) integrates the evolutionary concepts of irreversibility and uncontrolled technological change in an environmental I-O model.

Faber and Proops (1992) also consider the interactions between the environment and the economy from an evolutionary perspective. They state that the depletion of natural resources, and the invention of new production techniques may cause a shift in environmental-economic interactions. The emphasis in their treatment of innovation and invention is on economic development and the environment, unlike Schumpeter's theories on innovation, which focus on the creativity of the firm. The Neo-Austrian capital theory and models incorporate time in various dimensions such as irreversibility, growth and development (Faber *et al.*, 1989).

Aspects of models of technological change and economic evolution

Models in economic evolution are fundamentally different from the other model types discussed in this chapter as they are not based on optimization or equilibrium behaviour. Table 5.9 presents some characteristics of the models discussed in this section.

Models with technology, dynamics and evolution combined with material flows

Besides the concepts of material balance, irreversibility of extraction and the second law of thermodynamics, other concepts that are crucial for reducing or changing material flows are predictable and unpredictable technological change, evolution, sudden or abrupt changing of agents' preferences. In the model types discussed in Sections 5.2 to 5.5 only predictable technological and gradual preferential changes may be incorporated. In evolutionary models also unpredictable changes (novelties) and sudden preferential changes may be incorporated. These changes may reduce or shift the material flows and the problems associated with these flows. In most model types the environmental-economic relationship is mainly described from an economic perspective, implying that the environment is only modelled as providing the inputs and absorbing the outputs from the economic system.

The modelling of technological change and evolutionary modelling are new related approaches which are still mostly in the conceptual phase. Therefore, only a few theoretical and empirical models exist (Faber and Proops, 1992; Clark *et al.*, 1995; Allen, 1996b). At present, models with an evolutionary economic approach dealing explicitly with material flows do not exist. The evolutionary economic approach is a promising approach for dealing with non-gradual changes that may help to study policies and long-run developments affecting material flows. The most important insight from the approach with technological change and economic evolution for material flows is that non-deterministic changes and behaviour may be included. These changes may seriously influence the physical flows.

Table 5.9. The characteristics of models of technological change and economic evolution.

Goal of the model	Analysing the effects of evolutionary technological or preferential changes on the economy and environment
Type of model	Models in which non-mechanistic and non-deterministic behaviour and changes, learning, technological changes and evolution may be incorporated
Time aspect	Dynamic
Spatial scale	All scales possible
Aggregation level	All aggregation levels possible
Units	Prices, technologies, quantities

Endogenous growth models

In addition to the exogenous growth models described in Section 5.5, endogenous growth models including environmental issues are considered here, although endogenous growth models are based on the neoclassical, mechanistic approach, which is itself based on equilibria. Endogenous growth models emerged from exogenous growth models because the latter could not deal with endogenous technological change (see, for an extensive overview, Barro and Sala-i-Martin, 1995).

In endogenous growth models 'technical change and accumulation of technical knowledge are the result of economic decisions regarding investment in physical or human capital and R&D activities' (Gradus and Smulders, 1993). In endogenous growth models each production factor may be accumulated endogenously, whereas in exogenous growth models production factors grow at an exogenous rate.

Barro and Sala-i-Martin (1995) give an example of an endogenous growth model with labour-augmenting technological change. The labour will also depend on technological change, which results in $L(t)=L^*A(t)$. Then the change of capital over time becomes the following (compare with the change of capital in an exogenous growth model in Section 5.5):

$$\dot{k} = s\ f(k,A(t)/k\ -\ (n+\delta)$$

The output now also depends on the level of technology which may be changing over time.

Recently, many types of models with endogenous economic growth and natural resources have been studied (for an overview, see Barro and Sala-i-Martin, 1995; some examples are Van der Ploeg and Withagen, 1991; Gradus and Smulders, 1993; Eismont, 1994; Bovenberg and Smulders, 1995 and 1996; Huang and Cai, 1994; Bovenberg and De Mooij, 1996). These theoretical neoclassical studies may incorporate technological change, for example for pollution-saving, and the connection between environmental quality and growth. These models may be used to study how the government should intervene in order to obtain the optimal levels of growth and environmental quality.

The analytical (mainly optimizing) models with endogenous growth are embedded in neoclassical theory assuming, for example, that substitution between resources and capital is unlimited. Material flows studies require that resources and pollution are physically and thermodynamically (MB conditions) incorporated. Then, interesting questions concerning material flows may be analytically studied with these growth models.

5.7. Conclusions and Prospect

The goal of this chapter was to examine the potential use of various economic models for material flows modelling. The division of the various models into categories is a pragmatic choice because in the literature models are often a mixture of two or more model types. Table 5.10 gives a concise overview of the various models discussed in this chapter, summarizing Tables 5.1, 5.2, 5.7, 5.8 and 5.9.

Of course, none of the model types discussed in this chapter is perfect for all questions that may be asked. The choice of a model type should depend on the problem, the data availability and the goal of the study. In this chapter the main question was whether specific model types incorporate or refer to material flows. Table 5.11 lists whether and how the main aspects of material flow modelling can be included in the various model types. This table is based on the discussion of material flows modelling in various model types in Sections 5.2 to 5.6.

Economic models of natural resources may be appropriate for modelling material flows. These models are mainly used for allocating resources, but they may also be used to address pollution issues. The applications of these models mainly consider only one specific material, without taking into account potential substitutes.

Pollution models are optimization and equilibrium models with externalities in a neoclassical welfare-theoretic framework. These models are used to analyse externalities and the optimization of those externalities, for instance by implementing specific policies. In theoretical and applied pollution models, physical flows are not generally taken into account, because the models are in monetary terms. By including material flows and subsequently material balance conditions these models may address the physical effects of policies.

In contrast with pollution models, environmental input-output (I-O) models are mainly used for describing the relationship between economic sectors and the environment. They may also be combined with optimization modelling, for example to simulate policies. I-O models may also serve as a data-input for other model types, such as pollution or macroeconomic models.

Macroeconomic models in themselves do not seem very appropriate for studying material flows because of their high level of aggregation, but they may provide scenarios for studies using other, more disaggregated models. Models in which evolutionary changes, technology, evolution and dynamics are explicitly included are described as a separate model type because they may include unpredictable or non-mechanistic changes. These models are relatively new, and therefore not yet widely used. However, they seem to be potentially useful for modelling material flows: for example, because technological change may be included.

The theoretical physical flow and economic models discussed in Chapters 4 and 5 are the basis of the models applied to M-P chains in Part III of this study. M-P chain analysis aims to integrate the physical, environmental and economic aspects of material and product flows, and it allows important features such as substitution, recycling and reuse, and material balance conditions to be addressed in a single framework.

The physical flow models presented in Chapter 4 and the economic models in Chapter 5 are both applied in a range of studies in Part III. Table 5.12 shows which type of physical and economic models are combined in the applications. Material flow analysis and physical input-output analysis are conceptually the same type of physical flow model and therefore not distinguished here. The static optimization model of Chapter 6 combines a physical I-O model (or MFA) with a pollution model. In Section 6.5 a model with two technologies will be analysed. Chapter 7 will present a general equilibrium model based on physical I-O analysis and a pollution model with externalities. A dynamic model based on physical I-O analysis and parts of an economic model of natural resources will be given in Chapter 8. Chapter 9 will present a dynamic model that is a combination of LCA and an economic model of natural resources. The applied general equilibrium model of Chapter 10 combines a physical I-O model with a pollution model which has elements from monetary I-O and macroeconomic models.

Table 5.10. General characteristics of model types.

	Economic models of natural resources	*Pollution models*	*Input-output models*	*Macroeconomic models*	*Models of technological change and economic evolution*
Focus	Optimal allocation of natural resources over time	Optimal regulation of polluting activities	Sectoral structure; interaction, indirect and multiplier effect of changes in demand and production technology	Forecasting or scenario analysis of macro-impacts of environmental policies	Irreversible change, gradual and discrete jumps, co-evolution of environment and economy
Theoretical framework	Neoclassical; micro-economic	Neoclassical; micro-economic; externalities	Leontief production function for primary and intermediate inputs; meso-level	(Disequilibrium) relationships between macroeconomic variables	Non-mechanistic, disequilibrium, non-deterministic, learning
Temporal features	Dynamic	Static/dynamic	Mainly static	Dynamic	Dynamic
Model type	Optimization	Optimization; equilibrium	Descriptive (optimization)	Descriptive; forecasting	Descriptive

Table 5.11. Characteristics of model types with regard to material flows.

	Economic models of natural resources	*Pollution models*	*Input-output models*	*Macroeconomic models*	*Models of technological change and economic evolution*
Potential integration of material flows	Yes, in practice few models with extraction and pollution	Few examples	Yes, combination of physical and monetary integration rare	No	Rarely done, but possible
Question related to material flows	Optimal allocation of resources and pollution	Optimal regulation of extraction and pollution	Description of material flows between economic sectors and the environment	Macroeconomic impacts of macro-scenarios for materials-oriented policies	Analyses of potential long-run changes in material flows and their economic and environmental effects
Substitution of materials	Yes	Yes	No, because of fixed technological coefficients	No, models are too aggregated	Possible
Recycling of materials	Yes, but limited	Rarely done, but possible	Possible	Only exogenously	Few examples
Material balance	Possible	Not often done, but possible	Possible	No	Possible
Technological change	Yes, deterministic changes	Yes, deterministic changes	Exogenous change, no choice	Yes	Non-deterministic changes, evolutionary processes

Table 5.12. An overview of the physical and economic models that are used for modelling M-P chains.

	Static optimization model (Chapter 6)	*General equilibrium model (Chapter 7)*	*Dynamic simulation model (Chapter 8)*	*Dynamic simulation model (Chapter 9)*	*Applied general equilibrium model (Chapter 10)*
MFA/Physical I-O analysis	+	+	+		+
LCA				+	
Economic models of natural resources			+	+	
Pollution models	+	+			+
Economic I-O models					+
Macroeconomic models					+
Technological change and evolutionary models	+				

CHAPTER 6

A STATIC OPTIMIZATION MODEL FOR RAIN GUTTERS[1]

6.1. Introduction

Recently, integrated analysis of resource and pollution issues has gained interest as it can assist in understanding environmental problems associated with resource extraction, generation and emission of waste, or combinations of these. This has resulted in attention for material flows through economic systems, in order to deal simultaneously with processes of extraction, production, consumption, substitution between materials and other factors of production, recycling of materials and products, and waste treatment or disposal. The concept of material balance (MB) is relevant in this context because it requires equality of the total inflow and total outflow of materials in production, consumption, or even entire economic systems, in the absence of accumulation of material in economic stocks, products or capital (see Section 2.4 of Chapter 2). There are many studies that describe material flows through the economy using models that formalize the MB-concept. Examples are Ayres and Kneese (1969), Kneese *et al.* (1970), Ayres (1978), Gilbert and Feenstra (1992), Van den Bergh and Nijkamp (1994), Starreveld and Van Ierland (1994), and Weaver *et al.* (1995). Most of these studies focus on the flow of one or more specific materials and do not explicitly consider the link between the material flow and economic behaviour, processes and products. As with the study on the optimal recycling of plastics of Starreveld and Van Ierland (1994), here a static optimization model including the material balance principle is presented. This model optimizes the costs for the demand for a service which includes, besides the recycling of several materials, also demand and production functions.

An important concept for the analysis performed here is that of the material-product (M-P) chain (see Chapter 1). The demand for a service can often be met by a number of different products. Furthermore, products usually consist of more than one material. As a result, an M-P chain will generally include multiple products and multiple materials. Considering only a single material will generally provide an incomplete representation of the M-P chain associated with a particular product or service. However, taking into account all materials and products that, in one way or another, are linked to each other may turn out to be impossible for analytical or computational reasons, and furthermore not necessary for the problem or issue at hand. One may therefore restrict the analysis of empirical M-P chains to the most essential materials and products, referred to as truncated chains (Opschoor, 1994).

[1] This chapter is a slightly revised version of an article that appeared in *Environmental and Resource Economics* (Kandelaars and Van den Bergh, 1996a).

The actual demarcation of an M-P chain will thus depend on the specific objective of the analysis, which may relate to prudent use of a non-renewable resource, a shift from a toxic to a non-toxic material, etc.

The purpose of the analysis in this chapter is to explore how policies or strategies applied to different stages of an M-P chain differ in their impact on a number of indicators, including the use of materials and products, and the net costs of meeting demand for a particular service. The optimization models in this chapter maximize the total costs for consumers of meeting the demand for a service under a set of physical constraints. These total costs may include the costs or revenues of waste treatment after the products are disposed of. The model may be changed by imposing physical or economic policies on, for instance, products, recycling and waste treatment.

Section 6.2 addresses the issues of recycling, reuse and substitution which are of major importance to a treatment of M-P chains. Sections 6.3 to 6.5 describe in detail formal models of two types of M-P chains. Two-materials-one-product chains with exogenous and endogenous prices are discussed in Sections 6.3 and 6.4, respectively. Section 6.5 presents an M-P chain with two alternative technologies. In Section 6.6 an application of an M-P chain is presented that concerns the use of pvc and zinc in rain gutters. A final section presents conclusions.

6.2. A Model with Recycling, Reuse and Substitution

Recycling, reuse and substitution are relevant on the level of materials as well as products, and they can substantially affect the composition of M-P chains. Recycling of materials allows the use of one virgin material to decrease without increasing the use of another virgin material. Reuse of products allows the use of one product several times. Recycling may also require inputs such as energy, water and transportation. These have not been included in the model presented here. The main difference between the recycling of materials and the reuse of products is that in the latter case no transformation between products and materials is needed.

Three types of substitution can be distinguished: (1) between materials, (2) between materials and non-material inputs, and (3) between products. The first type can be regarded as resulting from a direct replacement of one material by another. In the second case where a material input is replaced by a non-material input, substitution may result from more efficient use of materials, e.g. through more working hours, or more generally from a change in the production technique. For addressing this type of substitution a production function with both types of inputs (i.e. materials and labour) is needed. It should be noted that the substitution between materials and non-materials is restricted by physical limits, because a physical product cannot be made with only labour or capital. Finally, the third type of substitution results from a change in the demand for products (see Section 3.4.2 of Chapter 3).

6.3. A Static Two-Materials-One-Product Chain with Exogenous Prices

In this section a model of a two-materials-one-product chain is presented. The exogenous elements are: the prices of production, recycling and dumping; the demand for the service; and the production techniques. The outputs of the model are the share of new and reused products, the percentages of recycled materials and the total net costs. This M-P chain is represented in Figure 6.1.[2]

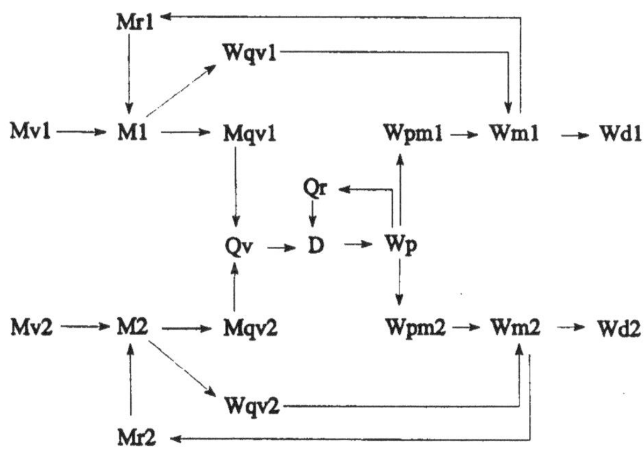

Figure 6.1. A two-materials-one-product chain.

The total input of materials of type i into the product consists of virgin materials, M_{vi}, and recycled materials, M_{ri}. These are substitutes. New products, Q_v, are produced from the material inputs M_1 and M_2. In the production process a part of the inputs ends up in the new products, M_{qvi}, while the remainder is production waste, W_{qvi}. This production waste consists of materials which are assumed to have the same quality as the materials that end up as waste material after use. The demand, D - which is equal to the supply - is satisfied by either new products, Q_v, or reused products, Q_r. When the products are used up, they become product waste, W_p (in units of product). Of this product waste one part is reused, Q_r, while the other part results in flows of waste material included in the product. The amount of each material in the product can be calculated by means of material balance conditions. Total waste material, W_{mi}, is the sum of the waste material of products, W_{pmi}, and the production waste, W_{qvi}. This waste will either be recycled, M_{ri}, or otherwise treated (for example, dumped), W_{di}. Note that all materials and products remain in the materials loop ($M_i \rightarrow W_{mi} \rightarrow M_{ri}$), the product loop ($D \rightarrow W_p \rightarrow Q_r$), or are

[2] In Appendix 6.1 a list of the symbols is given.

dumped (W_{di}).

The costs of meeting the demand are split up between non-material costs of new products (e.g. labour and capital costs per unit of output), $p_{qv}Q_v$, costs of the use of virgin and recycled materials, $p_{mvi}M_{vi}$ and $p_{mri}M_{ri}$, the costs of reused products, $p_{qr}Q_r$ and costs of waste dumping, $p_{wdi}W_{di}$. It is assumed that the production waste is added to the waste material without any costs and that the transformation of product waste into waste material (from W_p to W_{pmi}) is costless.

As an objective, the M-P chain aims to minimize the net costs of satisfying the demand, including waste treatment costs. Therefore, the objective function for this M-P chain becomes:

$$\min_{c_{rmi}, c_{rp}} \Omega = p_{qv}Q_v + p_{qr}Q_r + \sum_{i=1}^{2}[\, p_{mvi}\, M_{vi} + p_{mri}\, M_{ri} + p_{wdi}W_{di}] \tag{1}$$

Next, the system description consists of: (1) demand equations, (2) production conditions, (3) waste and recycling relationships, (4) constraints on the decision variables, and (5) derived MB conditions.

Demand

The demand is given exogenously.

$$D = \bar{D} \tag{2}$$

The demand is satisfied by new products, Q_v, and recycled products, Q_r.

$$Q_v + Q_r = D \tag{3}$$

Production

The number of new products, Q_v, is a function of two different material inputs, M_1 and M_2:

$$Q_v = f(M_1, M_2) \tag{4}$$

The input of materials in the new products consists of virgin materials, M_{vi}, and recycled materials, M_{ri}:

$$M_i = M_{ri} + M_{vi} \text{ for } i=1,2 \tag{5}$$

The input of materials in the production process, M_i, is equal to the sum of materials ending up in the product, M_{qvi}, and the production waste of materials i, W_{qvi}:

$$M_i = M_{qvi} + W_{qvi} \text{ for } i=1,2 \tag{6}$$

Waste and recycling
In this static setting the number of waste products is equal to the demand:

$$W_p = D \tag{7}$$

The number of reused products is a part, c_{rp}, of the product waste, W_p, where this relationship is assumed to be linear and constant:

$$Q_r = c_{rp}W_p \tag{8}$$

The amount of materials obtained out of product waste, W_{pmi}, equals the materials contents, m_i, of the products that are not recycled after they are used, W_p-Q_r, where the materials coefficient, m_i, represents the use of materials i per unit of the product.

$$W_{pmi} = m_i(W_p - Q_r) \text{ for } i=1,2 \tag{9}$$

$$m_i = \frac{M_i - W_{qvi}}{Q_v} \text{ for } i=1,2 \tag{10}$$

The production waste, W_{qvi}, is a function of the materials that go into the production process. Note that g_i should satisfy $M_i \geq W_{qvi}$:

$$W_{qvi} = g_i\,(M_i) \text{ for } i=1,2 \tag{11}$$

The total waste material, W_{mi}, is the sum of the production waste, W_{qvi}, and the non-recycled product waste in terms of materials, W_{pmi}:

$$W_{mi} = W_{qvi} + W_{pmi} \text{ for } i=1,2 \tag{12}$$

The quantity of recycled materials, M_{ri}, is a part, c_{rmi}, of the waste material, W_{mi}, where this relationship is assumed to be linear and constant.

$$M_{ri} = c_{rmi}W_{mi} \text{ for } i=1,2 \tag{13}$$

The quantity of waste material which is treated, W_{di}, is equal to the part of the total waste material that is not recycled:

$$W_{di} = W_{mi} - M_{ri} \text{ for } i=1,2 \tag{14}$$

Decision variables
The coefficient c_{rmi} represents the part of the materials of type i that is recycled (i=1,2). c_{rp} represents the part of the product waste that is reused. The values of the coefficients c_{rmi} and c_{rp} are between 0 and 1, as the maximal percentage that can be recycled or reused is 100%:

$$0 \leq c_{rmi} \, , \, c_{rp} \leq 1 \text{ for } i=1,2 \tag{15}$$

The coefficients c_{rmi} and c_{rp} are decision variables in the model. Their optimal value will be determined by solving the optimization model. These coefficients may also be restricted by technological, economic or institutional constraints. This will not be considered here.

Derived MB conditions

The material balance restrictions of the above model allow the derivation of three other MB conditions:

- On the product level[3]: $W_{pmi} \overset{(9)}{=} m_i(W_p-Q_r) \overset{(7)}{=} m_i(D-Q_r) \overset{(3)}{=} m_iQ_v \overset{(10)}{=} M_i-W_{qvi} \overset{(6)}{=} M_{qvi}$. The waste material of non-recycled product waste equals the material inputs in products.
- On the production level: $W_{mi} \overset{(12)}{=} W_{qvi}+W_{pmi} = W_{qvi}+M_{qvi} \overset{(6)}{=} M_i$, using the result of (i) in the second step. The waste material is equal to the inputs of materials.
- On the system level: $M_{vi} \overset{(5)}{=} M_i-M_{ri} = W_{mi}-M_{ri} \overset{(14)}{=} W_{di}$, using the result of (ii) in the second step. The input of virgin materials equals the waste material treated (for an overview of material balance conditions in the present model see Table 6.2).

The model can be rewritten by substituting restrictions into the objective function. The model then becomes:

$$\min_{c_{rmi},c_{rp},M_i} \Omega = p_{qv}(1-c_{rp})\bar{D} + p_{qr}c_{rp}\bar{D} + \sum_{i=1}^{2}[\,(p_{mvi}+p_{wdi})(1-c_{rmi})+p_{mri}c_{rmi}]\,M_i \tag{16}$$

with:

$$(1-c_{rp})\bar{D} = f\,(M_1,M_2) \tag{17}$$

and the conditions of (15). This model has five decision variables: c_{rm1}, c_{rm2}, c_{rp}, M_1 and M_2.

The model of (16) and (17) is as follows: if the price of recycling materials is smaller than the sum of the price of virgin materials and treatment, i.e. $p_{mri} < p_{mvi} + p_{wdi}$, then the maximum amount of waste material of type i is recycled. This implies that it is optimal to recycle all waste material i, i.e. $c_{rmi}^*=1$.[4] Otherwise, the optimal $c_{rmi}^*=0$. If $p_{mri}=p_{mvi}+p_{wdi}$ then the amount of materials of type i that is recycled does not have an impact on the total costs. If the costs of product recycling are lower than the sum of the costs of a new product, materials recycling and

---

[3] The numbers of the equations used are indicated between brackets.

[4] Asterisks are used to indicate optimal levels.

treatment, then the waste of products is recycled. With a linear production function the optimal c_{rp} will thus be one of two extremes, i.e. $c_{rp}^*=0$ or $c_{rp}^*=1$. With a non-linear production function the optimal values can relate to an interior solution. The values of the materials variables M_1 and M_2 depend on the demand and on the production function f. Note that in this model the prices of materials and products are fixed.

6.4. Endogenous Price of Reuse

In the model of Section 6.3 all prices were exogenous. In this section the price of recycled products is assumed to increase with the number of products that are recycled, incorporated via quadratic specification $p_{qr}=Q_r^2$. Using $Q_r=c_{rp}D$ (from equations 7 and 8) the minimization function with the substitution of restrictions now becomes:

$$\min_{c_{rmi},c_{rp},M_i} \Omega = p_{qv}(1-c_{rp})\bar{D}+(c_{rp}\bar{D})^2\, c_{rp}\bar{D}+\sum_{i=1}^{2} [(p_{mvi}+p_{wdi})(1-c_{rmi})+p_{mri}c_{rmi}]\, M_i \quad (18)$$

which is similar to objective (16), except for the second term. The restrictions remain the same, i.e. equations (15) and (17). The optimal recycling parameters and the minimal costs can be determined by partially differentiating Ω to c_{rp} which results in:

$$c_{rp} = \sqrt{\frac{p_{qv}}{3\bar{D}^2}} \quad (19)$$

For example, when the demand is equal to 3 (D=3) and the price of new products is equal to 12 ($p_{qv}=12$) the optimal reuse percentage is approximately 67% ($c_{rp}^*=2/3$). The optimal recycling percentages are equal to 0 or 1, as in Section 6.3. A consequence of endogenous prices is that the optimal shares of materials and product recycling are no longer necessarily corner solutions, because the optimization model is non-linear. The optimal levels of recycling and costs can be calculated for an M-P chain with endogenously determined prices if the price functions are known and the first derivatives of these functions exist.

6.5. Two Production Technologies in an M-P Chain

A second interesting extension of the M-P chain is where a service is considered that can be provided by two alternative production technologies. The graphical representation is shown in Figure 6.2.

The demand is satisfied by products which are made by two alternative technologies, i.e. product Q_i is made by technology i with i=1,2. Products Q_1 and Q_2 are perfect substitutes of each other. The total product costs are: (1) the non-materials costs of new products, $p_{qvj}Q_{vj}$ and the materials costs of a new product which are split up between the costs of virgin and recycled materials, $p_{mvi}M_{vi}$ and $p_{mri}M_{ri}$, (2) the costs of reused products, $p_{qrj}Q_{rj}$ and (3) the costs of waste treatment, $p_{wdi}W_{di}$. \

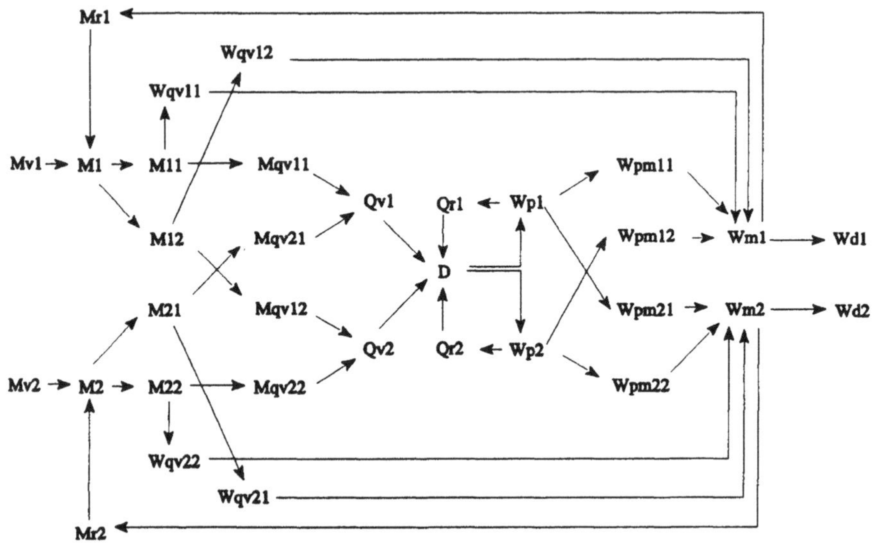

Figure 6.2. A two-materials-one-product chain with two technologies.

The objective function can then be formulated as:

$$\min_{c_{rmi}, c_{rpj}, M_{ij}, Q_j} \Omega = \sum_{i=1}^{2} \sum_{j=1}^{2} p_{qvj}Q_{vj} + p_{qrj}Q_{rj} + p_{mvi} M_{vi} + p_{mri} M_{ri} + p_{wdi}W_{di} \tag{20}$$

The restrictions on the demand side, the production side and the waste and recycling side are given in Table 6.1 and are analogous to those discussed in the previous two sections. After substituting the model conditions the objective function becomes:

$$\min_{c_{rmi}, c_{rpj}, M_{ij}, Q_j} \Omega = \sum_{i=1}^{2} \sum_{j=1}^{2} p_{qvj}(1-c_{rpj})Q_j + p_{qrj}c_{rpj}Q_j + [(p_{mvi}+p_{wdi})(1-c_{rmi})+p_{mri}c_{rmi}]M_{ij} \tag{21}$$

with the following conditions:

$$(1-c_{rpj})Q_j = f_j\,(M_{1j}, M_{2j}) \text{ for } j=1,2 \tag{22}$$

$$\sum_{j=1}^{2} Q_j = \bar{D} \tag{23}$$

$$0 \leq c_{rmi}\ ,\ c_{crpj} \leq 1 \text{ for } i,j=1,2 \tag{24}$$

Table 6.1. Overview of the models of the two-materials-one-product chain and the two technologies M-P chain.

	Two-materials-one-product chain (for i=1,2)	Two technologies (for i,j=1,2)
Demand side	$D=\bar{D}$ $Q_v+Q_r=D$ -	$D=\bar{D}$ $Q_{vj}+Q_{rj}=Q_j$ $\sum_{j=1}^{2} Q_j = D$
Production side	$Q_v= f\ (M_1,\ M_2\ )$ $M_i=M_{ri}+M_{vi}$ -	$Q_{vj}= f_j\ (M_{1j},\ M_{2j}\ )$ $M_i=M_{ri}+M_{vi}$ $M_i = \sum_{j=1}^{2} M_{ij}$
Waste and recycling side	$M_i=W_{qvi}+M_{qvi}$ $W_p=D$ $Q_r=c_{rp}W_p$ $W_{pmi}=m_i(W_p-Q_{rj})$ $m_i=\frac{M_i-W_{qvi}}{Q_v}$ $W_{qvi}=g_i(M_i)$ $W_{mi}=W_{qvi}+W_{pmi}$ $M_{ri}=c_{rmi}W_{mi}$ $W_{di}=W_{mi}-M_{ri}$	$M_i=\sum_{j=1}^{2} W_{qvij}+M_{qvij}$ $W_{pj}=Q_j$ $Q_{rj}=c_{rpj}W_{pj}$ $W_{pmij}=m_{ij}(W_{pj}-Q_{rj})$ $m_{ij}=\frac{M_{ij}-W_{qvij}}{Q_{vj}}$ $W_{qvij}=g_{ij}(M_{ij})$ $W_{mi}=\sum_{j=1}^{2} W_{qvij}+W_{pmij}$ $M_{ri}=c_{rmi}W_{mi}$ $W_{di}=W_{mi}-M_{ri}$

The decision variables are c_{rpj}, c_{rmi}, M_{ij}, Q_j for i,j=1,2. The optimal values for c_{rpj} and c_{rmi} are calculated using the Lagrange function, after which the optimal levels of the inputs of materials i for product j, M_{ij}, the quantity of products of type j made to satisfy the demand, Q_j, and the net costs are determined. In Appendix 6.2 it is shown that the optimal solutions for the c_{rmi} are extremes if the prices for new and recycled materials/reused products are not the same. Intuitively, the maximum amount of materials will be recycled if recycled materials are cheaper than new materials. The percentages of reused products (c_{rpj}), material inputs (M_{ij}), and the products (Q_j) to meet the demand depend on the production functions and the demand as indicated in Appendix 6.2. The main difference with the previous models (Sections 6.3 and 6.4) is that here a choice can be made between products made with different production technologies. Tables 6.2 and 6.3 give overviews of the MB conditions and the decision variables in the M-P chains described in Sections 6.3 and 6.5.

Table 6.2. Overview of the material balance conditions for the two M-P chains.

	Two-materials-one-product chain (for i=1,2)	Two technologies (for i,j=1,2)
Product level	$M_{qvi}=W_{pmi}$	$M_{qvij}=W_{pmij}$
Production level	$M_i=W_{mi}$	$M_i=W_{mi}$
Chain level	$M_{vi}=W_{di}$	$M_{vi}=W_{di}$

Table 6.3. Overview of the decision variables of the two M-P chains.

	Two-materials-one-product chain (for i=1,2)	Two technologies (for i,j=1,2)
Recycling level	c_{rmi}, c_{rp}	c_{rmi}, c_{rpj}
Material inputs level	ratio between M_1 and M_2	ratio between M_{1j} and M_{2j}
Product level	-	ratio between Q_1 and Q_2

6.6. Application for Zinc and Pvc Rain Gutters

In this section an application of M-P chain-analysis is given to illustrate the approaches adopted in previous sections. The model of Section 6.3 is applied to the case of pvc and zinc gutters. The focus of the M-P chain in this section is the demand for gutters. Two types of gutters will be compared, i.e. zinc and pvc. Other types of gutter, e.g. aluminium, polyester and wood, are not taken into account here, as they provide only a small part of the total market for gutters.

The M-P chain includes:

- three basic materials, i.e. zinc ore, iron ore and pvc;[5]
- two intermediate materials, i.e. zinc and galvanized steel;
- two intermediate products, i.e. two types of fastening-pieces; and,
- two final products, i.e. complete zinc and pvc gutters.

The two final products provide the same service, and are perfect substitutes. In other words, they satisfy a single demand for services. The two types of fastening pieces both consist of galvanized steel, but are different in weight for pvc or zinc gutters. Water pipes are not taken into account because they can be combined with each type of rain gutter. This M-P chain is illustrated graphically in Figure 6.3. In principle the zinc and pvc gutters and the fastening-pieces of galvanized steel can be recycled as both products and as materials. Recycling of gutters is not practised at the moment because of legal regulations, technical and economic restrictions.

For the application to gutters, it is assumed that there is no product reuse and the production functions include a single input. Data is obtained from Fraanje and Verkuijlen (1996), Gorter (1994) and Tauw Milieu (1994) and describe the M-P

[5] Pvc is regarded here as a basic material like zinc and iron ore. Physically, this is not correct because pvc is a produced compound made of several materials. Conceptually, however, it makes no difference whether a material is a basic ore or produced, because, by definition, any material in the model is exogenous.

chain for gutters in the Netherlands. The static chain optimization model consists of the following objective function:[6]

$$\min_{Q_{zg},Q_{pg}} \Omega = Q_{zg} * p_{q,zg} + Q_{pg} * p_{q,pg} - M_{r,z} * p_{mr,z} - M_{r,p} * p_{mr,p} - M_{r,gs} * p_{mr,gs} \quad (25)$$

The objective is to minimize the net costs of satisfying the demand: the costs of the zinc gutters (Q_{zg}*$p_{q,zg}$) and the costs of the pvc gutters (Q_{pg}*$p_{q,pg}$) minus the revenues for the recycled materials ($M_{r,i}$*$p_{mr,i}$ for i=zinc,pvc, galvanized steel). The costs of waste treatment of the gutters that are disposed of is assumed to be zero.[7] In the objective function the revenues of recycled materials are explicitly taken into account in order to affect the choice of consumers for a certain type of gutter. Thus, although the recycling percentage for a type of material is fixed, the decision concerning which type of gutter is directly related to the recycling of the materials embodied in that gutter.

As in the foregoing sections, the restrictions are subdivided in categories related to (1) demand, (2) production, and (3) waste and recycling. These restrictions can be compared with those in Table 6.1. The static optimization model described has eight variables, i.e. five price variables and three recycling percentages. The decision variables are the division of the demand over the two products (see equation 27).

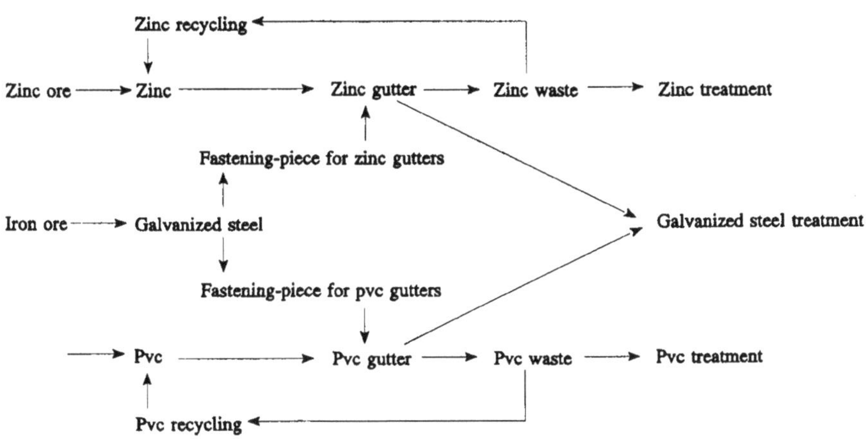

Figure 6.3. An M-P chain for gutters.

[6] An explanation of the symbols is given in Appendix 6.3.

[7] An interpretation is that the extra (variable) costs of waste gutters to the total costs of waste treatment (fixed) are zero.

Demand

The demand for standardized 12-metre gutters equals the number of gutters needed for new houses, i.e. 75,000, and for the renovation of houses, i.e. 54,000, which makes the total demand for gutters equal to 129,000. There are no technical constraints to replace one type of gutter with another. Hence,

$$\bar{D} = 129{,}000 \tag{26}$$

The demand for gutter services will be satisfied either by zinc or by pvc gutter services:

$$D = Q_{zg} + Q_{pg} \tag{27}$$

Production

For each gutter of type j that is used, it is also necessary to use one unit of fastening-piece:

$$Q_{fastj} = Q_{jg} \text{ for } j = zg, pg \tag{28}$$

The production of one kilogram of zinc (galvanized steel) requires $x_{z,z\text{-}ore} = 22.3$ ($x_{gs,z\text{-}ore} = 1.67$) kilograms of zinc ore and $x_{z,fe\text{-}ore} = 0$ ($x_{gs,fe\text{-}ore} = 1.44$) kilograms of iron ore:

$$M_{v,i} = x_{i,k\text{-}ore} * I_{k\text{-}ore} \text{ for } i = z, gs \text{ and } k = z, fe \tag{29}$$

The following technological production functions convert materials measured in kilograms to products measured in functional units. A standardized zinc (pvc) gutter of 12 metres consists of $m_z = 29.6$ ($m_p = 12.1$) kilograms of zinc (pvc) (compare with materials coefficient m_i in Section 3):

$$Q_j = \frac{1}{m_i} * M_i \text{ for } j = zg, pg \text{ and } i = z, p \tag{30}$$

A fastening-piece for zinc (pvc) gutters consists of $b_{zg} = 25.8$ ($b_{pg} = 16.8$) kilograms of galvanized steel:

$$Q_{fastj} = \frac{1}{b_j} * M_{gs} \text{ for } j = zg, pg \tag{31}$$

Note that the amount of galvanized steel needed for the fastening-piece of a pvc gutter service differs from the amount needed for the fastening-piece of a zinc gutter.

Waste and recycling

After use the gutters are not reused as such, which means that there are no reused products, but after converting the used products into materials, the recycling of these materials is possible. The conversions from products into waste material are described by (32) and (33). The amount of used zinc (pvc) depends on the number of zinc (pvc) gutters used. A zinc (pvc) gutter consists of m_z=29.6 kilograms of zinc (m_p=12.1 kilograms of pvc) which means that the total amount of used zinc (pvc) gutters has to be multiplied by m_z=29.6 (m_p=12.1) to obtain the total amount of waste from zinc (pvc) gutters:

$$W_{m,i} = m_i * Q_j \text{ for } i=z,p \text{ and } j=zg,pg \tag{32}$$

The amount of used galvanized steel depends on the amount of pvc and zinc gutter services used. A fastening-piece for zinc (pvc) gutters consists of b_{zg}=25.8 (b_{pg}=1-6.8) kilograms of galvanized steel. The total sum of waste of galvanized steel is therefore the sum of the waste of both types of fastening-pieces:

$$W_{m,gs} = b_{zg} * Q_{fastzg} + b_{pg} * Q_{fastpg} \tag{33}$$

The final set of equations shows the parts of the waste material which are recycled. The parts of the waste which are recycled depend on the prices of collection, recycling and treatment and on the government policy on recycling, e.g. a penalty for non-recycling.

A part, c_i, of waste material i is recycled:

$$M_{r,i} = c_i * W_{m,i} \text{ for } i=z,p,gs \tag{34}$$

The zinc, pvc and galvanized steel waste that are not recycled are treated:

$$W_{d,i} = W_{m,i} - M_{r,i} \text{ for } i=z,p,gs \tag{35}$$

This model is used to calculate the optimal distribution of the demand for gutters under different scenarios. Five scenarios will be developed for analysing the influences of technological and economic developments on M-P chains, by means of a static optimization model. The scenarios focus specifically on policies affecting the economic determinants of the flow of materials and products. The goal of the policies is to use less zinc because of its harmful non-point pollution; it is assumed that zinc is environmentally more harmful than pvc.

The following scenarios are incorporated in the static model:

1. base scenario;
2. product charge scenario;
3. recycling scenario;
4. product charge and collection subsidy scenario; and,
5. waste treatment scenario.

The base scenario, in which all variables are set at their 1990 level, serves as a reference scenario for the other scenarios in which one or more exogenous variables are changed. Hereafter, the different scenarios are elaborated and then the results are presented and discussed. Table 6.4 lists the control variables for each of the five scenarios.

1. Base scenario

In the base scenario the objective function is as follows (see equation 25):

$$\min_{Q_{zg},Q_{pg}} \Omega = Q_{zg} * 180 + Q_{pg} * 180 - M_{r,z} * 1 - M_{r,p} * 0 - M_{r,gs} * 0 \tag{36}$$

The price for a functional unit of 12 metres of gutter is Dfl 180 for both zinc gutters and pvc gutters. The revenues for the recycled materials are in guilders per kilogram materials. In 1990 one kilogram of used zinc cost Dfl 1. Used pvc does not have a positive value because it is cheaper to produce pvc from new materials than from recycled materials. Galvanized steel is not recycled for technological and economic reasons. In this static model the total costs for the consumers of the gutters equal the prices paid for the zinc and pvc gutters minus the revenues they receive from a recycling activity for the waste material of the used gutters.

The recycling percentages for used zinc, pvc and galvanized steel are 90%, 70% and 0%, respectively. These percentages are exogenous in this model, because they are determined by the recycling activity and not by the consumers. Used zinc is recycled in large quantities because it is easy to collect and it provides a positive revenue. Because of leaching and additions of other materials it is not possible to recycle all the zinc used in zinc gutters. The exact amount of leaching of zinc gutters is not known (Gorter, 1994). The reason for collecting used pvc may be more for 'green consciousness' than for economic reasons, as it is generally not profitable to do so. Thus, the consumers do not have a financial incentive to collect use pvc.

Table 6.4. Settings of control variables for each scenarios.

Scenarios / Control variables	Base	Product charge	Recycling	Product charge and collection subsidy	Waste treatment
Price zinc gutter (Dfl/unit)	180	180+30		180+20	
Price pvc gutter (Dfl/unit)	180				
Revenue from used zinc (Dfl/kg)	1			1.5	
Revenue from used pvc (Dfl/kg)	0			2	
Price of dumping galvanized steel (Dfl/kg)	0				$1+(Q_{zg}/200{,}000)$
Part of zinc recycled	0.9		0.95	0.95	
Part of pvc recycled	0.7		0.9	0.9	
Part of galvanized steel recycled	0		0.5		

2. Product charge scenario

In the product charge scenario the price of zinc gutters is changed by a government levy of Dfl 30 on the use of zinc gutters, making $p_{q,zg}=180+30$. The other prices and recycling percentages are taken to be equal to those in the base scenario.

3. Recycling scenario
In the recycling scenario the percentages of zinc, pvc and galvanized steel that are recycled are changed compared with the base scenario. The percentages (c_i, see equation 34) are changed from 90%, 70% and 0% to 95%, 90% and 50% for zinc, pvc and galvanized steel, respectively. This exogenous increase in recycling percentages may result from a government policy, such as a deposit-refund policy, or a covenant of the government with demolishers and renovators of houses.

4. Product charge and collection subsidy scenario
In this scenario two policies are used simultaneously: (i) a subsidy is given on both used zinc (Dfl 0.5) and pvc (Dfl 2); and, (ii) a charge of Dfl 20 is put on the use of zinc gutters. The increases in the prices of used zinc and pvc have an impact on the recycling percentages and the net costs of satisfying the demand. It is assumed that the percentages of recycling rise to 95% (70%) when the price of used zinc (pvc) rises to Dfl 1.5 (2) per kilogram. A product charge of Dfl 20 per zinc gutter implies that zinc gutters become relatively less attractive financially.

5. Waste treatment scenario
In this scenario the government imposes costs on the treatment of galvanized steel. The goal is to decrease the use of zinc gutters which use more kilograms of galvanized steel per unit. The price of the treatment of galvanized steel is therefore made dependent on the number of zinc gutters consumed, i.e. $p_{wd,gs}=f(Q_{zg}) = a+b*Q_{zg}$, with constants $a=1$ and $b=1/200{,}000$. The objective function is then:

$$\min_{Q_{zg},Q_{pg}} \Omega = Q_{zg} * 180 + Q_{pg} * 180 - M_{r,z} * 1 + W_{d,gs} * \left(1 + \frac{Q_{zg}}{200{,}000}\right) \tag{37}$$

To compare the five scenarios and the current situation, some relevant performance indicators are presented in Table 6.5. These variables are: the net costs of satisfying the demand, and the allocation of the demand over the zinc and pvc gutters.

Table 6.5. Performance of indicators under different scenarios.

Scenarios / Indicators	Base	Product charge	Recycling	Product charge and collection subsidy	Waste treatment	Current situation
Net costs (in millions of Dfl)	19.8	23.2	19.6	21.8	20.1	20.5
Zinc gutters (in units)	129,000	0	129,000	0	75,600	103,200
Pvc gutters (in units)	0	129,000	0	129,000	53,400	25,800

In the first four scenarios the Lagrangian optimization model is linear in both the objective function and the restrictions. The optimal solution leaps from one extreme to another at certain levels of the control variables. In the fifth scenario the optimal allocation can be an interior solution, because $p_{wd,gs}$ is now a function of Q_{zg} which makes the Lagrangian non-linear. The current situation is described to give an idea of the situation in 1990. Note that in this model the possible external costs are not taken into account.

1. Results of base scenario and the current situation

In the base scenario the prices of zinc and pvc gutters are equal while the price of used zinc is positive and the prices of used pvc and galvanized steel are zero. Therefore, it is optimal to meet the demand for gutters only with zinc gutters. The net costs will be Dfl 19.8 million while in the current situation the net costs are Dfl 20.5 million. Compared with the optimal solution the current situation is not optimal which means that the consumers are not optimizing the net costs of their demand. The reason for this might be that they are not aware of all costs that occur, especially the revenues arising after the gutter is demolished.

2. Results of product charge scenario

In the product charge scenario the price of zinc gutters is increased by a levy in order to decrease the use of zinc. The net costs of one zinc gutter and one pvc gutter will be equal when the levy is equal to Dfl 26.64 per unit of zinc gutter. This means that when the levy is Dfl 26.64 per unit of zinc gutter it does not matter whether the zinc or the pvc gutter is chosen. When the levy is higher than Dfl 26.64, it is optimal to meet the demand for gutters only by pvc gutters and when it is lower it is still optimal to choose zinc gutters only, as in the case where no levy was imposed. Here the levy is Dfl 30 per functional unit making the net costs 23.2 million.

3. Results of recycling scenario

In the recycling scenario the recycling percentages for the three materials have increased. The optimal solution will be to choose only zinc gutters because used zinc is the only material which has a positive price, i.e. a negative effect on the net costs. The net costs will be Dfl 19.6 million. The effect on the net costs is modest, because the increase in the amount of used zinc is small, especially in comparison with the costs of purchasing and the revenues from the used zinc. The effect on the environment is significant because the recycling percentage of zinc increases, which decreases the use of non-recycled materials.

4. Results of product charge and collection subsidy scenario

In this scenario subsidies are given for the collection of used zinc and used pvc and a product charge is imposed on zinc gutters. An advantage of subsidizing used materials is that more materials are recycled, which has a positive effect on the environment. A product charge makes the zinc gutters less attractive. The optimal solution is that the demand is totally allocated to pvc gutters, with net costs of Dfl 21.8 million.

5. Results of the waste treatment scenario
The government is assumed to set $p_{wd,gs}=1+Q_{zg}/200{,}000$, so that the minimization function in terms of Q_{zg} and Q_{pg} equals:

$$\min_{Q_{zg},Q_{pg}} \Omega=180(Q_{zg}+Q_{pg})-0.9*29.6*Q_{zg}+(25.8*Q_{zg}+16.8*Q_{pg})(1+\frac{Q_{zg}}{200{,}000}) \quad (38)$$

Replace Q_{pg} by 129,000-Q_{zg} and differentiate to Q_{zg} to obtain the optimal Q_{zg}^*= 75,600 and the optimal Q_{pg}^*=53,400 with net costs of Dfl 20.1 million. In this optimization model the objective function is non-linear, which implies that the optimum is not necessarily one of the two extreme allocations as in the other 4 scenarios. The environmental effect is that less zinc and more pvc is used than in the base scenario.

6.7. Conclusions

This chapter has discussed the application of optimizing models for the theoretical and empirical analysis of material-product (M-P) chains. The chapter started with clarifying the concepts of recycling, reuse and substitution. Next, three types of static M-P chains were analysed. First, a two-materials-one-product chain with exogenous prices was studied, which showed that the optimal recycling percentages are corner solutions. Second, when the M-P chain is extended with endogenous prices the optimal solution is now not necessarily a corner solution. Third, in the M-P chain for two technologies the recycling percentages are corner solutions, while the material inputs and the demand for products depend highly on the shape of the production functions. The models include material balance conditions on each level. Therefore, the materials embodied in a product need to be recorded to obtain the amount of material waste. If various production technologies may be used the recorded material waste of the products is added to obtain the total material waste. It is an important insight that the inputs in the production function is relevant for the waste treatment stage of the products.

An application considered the impact of control variables in an M-P chain optimization model for gutters. The model minimizes the costs for consumers (e.g. construction firms or households) to meet their demand for gutters. These costs include the purchase price for the gutters and the revenues for the consumers of the waste material from gutters that are disposed of. The model shows that the current situation is financially not an optimal one for the consumers. It shows that policies may have a significant impact on the distribution of demand, the extraction of materials and the net costs of the M-P chain. The model may demonstrate to consumers that when choosing a certain product not only is the purchase price relevant, but also the revenues after use. Furthermore, the modest model shows that considering an M-P chain, including material balance conditions, with various products is a tool to analyse possible policy implementations.

The approach here may be extended by expanding production functions with non-material inputs and by the inclusion of dynamic relationships. A dynamic model would, however, be more complicated, because the accumulation of materials and products in the chain causes heterogeneity of products made in different years. These complications can be faced by adopting a vintage approach (see Chapters 8 and 9). This would be especially relevant for studying the evolution of one product into other types of products, each characterized by a specific material composition.

Appendix 6.1. List of symbols in Sections 6.3 to 6.5

Indices:

- i = material i
- j = product j
- ij = material i in/used for product j

Physical quantities (in kilograms):

- M_{vi} = virgin material input
- M_{ri} = recycled material input
- M_i = total material input
- W_{pj} = product waste
- W_{mi} = waste material
- W_{di} = treatment of waste
- W_{pmi} = waste material from transformed product waste
- W_{qvi} = production waste
- M_{qvi} = part of the material input in products

Physical quantities (in functional units):

- Q_{vj} = new products
- Q_{rj} = recycled products
- D = total demand
- Q_j = total of product j for $j=1,2$

Prices of physical quantities (in guilders per kilogram):

- p_{mvi} = virgin material input
- p_{mri} = recycled material input
- p_{wdi} = treatment of waste

Prices of physical quantities (in guilders per functional unit):

- p_{qvj} = new products
- p_{qrj} = recycled products

Coefficients:

- c_{rmi} = part of the waste material that is recycled
- c_{rpj} = part of the product waste that is recycled
- m_i = materials coefficient

Function

- Ω = net costs of satisfying the demand

Appendix 6.2. Lagrange conditions for the two-technologies M-P chain

$$\left.\begin{array}{c} L(c_{rmi},c_{rpj},M_{ij},Q_j) = \\ \sum_{i=1}^{2}\sum_{j=1}^{2} p_{qvj}(1-c_{rpj})Q_j + p_{qrj}c_{rpj}Q_j + [(p_{mvi}+p_{wdi})(1-c_{rmi})+p_{mri}c_{rmi}]\ M_{ij} - \\ \lambda_{ij}[(1-c_{rpj})Q_j - f_j(M_{1j},M_{2j})] - \\ \lambda_2(Q_1 + Q_2 - D) - \lambda_{3i}(c_{rmi}) - \\ \lambda_{4i}(c_{rmi} - 1) - \lambda_{5j}(c_{crpj}) - \lambda_{6j}(c_{rmi} - 1) \end{array}\right\}$$

Partially differentiating L to c_{rmi}, c_{rpj}, M_{ij} and Q_j for i,j=1,2 gives that:

$$\sum_{j=1}^{2}(-p_{mvi}-p_{wdi}+p_{mri})M_{ij}-\lambda_{3i}-\lambda_{4i}=0 \text{ for } i=1,2$$

$$-p_{qvj}Q_j+p_{qr}Q_j+\lambda_{1j}Q_j-\lambda_{5j}-\lambda_{6j}=0 \text{ for } j=1,2$$

$$[(p_{mvi}+p_{wdi})(1-c_{rmi})+p_{mri}c_{rmi}]+\lambda_{1j}\frac{\delta f_j}{\delta M_{ij}}=0 \text{ for } i,j=1,2$$

$$p_{qv}(1-c_{rpj})+p_{qr}c_{rpj}-\lambda_{1j}(1-c_{rpj})-\lambda_2=0 \text{ for } j=1,2$$

and $\lambda_{3i}c_{rmi}=0$, $\lambda_{4i}(c_{rmi}-1)=0$, $\lambda_{5j}c_{rpj}=0$ and $\lambda_{6j}(c_{prj}-1)=0$ for i,j=1,2, and equations 22 and 23 of Section 5.

(i) if $p_{mvi}+p_{wdi}<p_{mri}$ then $c_{rmi}=0$ (or $M_{ij}=0$)
(ii) if $p_{mvi}+p_{wdi}>p_{mri}$ then $c_{rmi}=1$ (or $M_{ij}=0$)
(iii) if $p_{mvi}+p_{wdi}=p_{mri}$ then $0\leq c_{rmi}\leq 1$

Then redefine $a_i=(p_{mvi}+p_{wdi})(1-c_{rmi})+p_{rmi}c_{rmi}$ for i=1,2
From equation 42 it follows that: $\lambda_{1i}=-a_i/(\delta f_j/\delta M_{1j})=-a_i/(\delta f_j/\delta M_{2j})$ for i,j=1,2
In the optimal solution: $\delta f_j/\delta M_{1j}=\delta f_j/\delta M_{2j}$ for j=1,2, after which λ_{1j} can be calculated by substituting a_i.

From equation 43 it follows that by substituting the value of λ_{1j} for j=1,2:
$p_{qv1}(1-c_{rp1})+p_{qr1}c_{rp1}-\lambda_{11}(1-c_{rp1})=p_{qv2}(1-c_{rp2})+p_{qr2}c_{rp2}-\lambda_{12}(1-c_{rp2})$ which gives a relationship between the c_{rp1} and c_{rp2}, after which the optimal M_{ij} and Q_j can be calculated with the use of equations 22 and 23.

Appendix 6.3. List of symbols in Section 6.6

Indices:

k-ore = z-ore for zinc ore, fe-ore for iron ore
i = z for zinc, p for pvc, gs for galvanized steel
j = zg for zinc gutters, pg for pvc gutters

Physical quantities (in kilograms):

$I_{k\text{-}ore}$ = inputs k-ore
M_i = total material input i
$M_{v,i}$ = virgin material input i
$M_{r,i}$ = recycled material input i
$W_{m,i}$ = waste material i
$W_{d,i}$ = dumped waste material i

Physical quantities (in functional units):

Q_j = gutter type j
$Q_{fast,j}$ = fastening-piece for gutter type j
D = demand for gutters

Prices (in guilders):

$p_{q,j}$ = price of one functional unit of a gutter of type j
$p_{mr,i}$ = price of one kilogram of recycled materials of type i
$p_{wd,gs}$ = price of the treatment of one kilogram of galvanized steel

Coefficients:

c_i = part of the waste material that is recycled
$x_{i,k\text{-}ore}$ = the amount of kilograms needed of k-ore to make one kilogram of material i
m_i = amount of kilograms of material i in gutter made of that material
b_j = amount of kilograms of galvanized steel needed for a fastening-piece of gutter j

Function

Ω = net costs of satisfying the demand

CHAPTER 7

A STATIC GENERAL EQUILIBRIUM ANALYSIS OF AN M-P CHAIN[1]

7.1. Introduction

Although it is often stated that resource extraction and environmental pollution are related via material flows through the economy, most studies in environmental economics do not treat this relationship explicitly. Of course, it could be argued that in the short run extraction and pollution can be treated separately. However, over a longer period of time the relationship between the two is extremely important. This is also recognized in the literature on economic growth and the environment, environmental sustainability, the de-linking of income and environmental pressure (the environmental Kuznets curve debate), and dematerialization and industrial metabolism (e.g. Ayres, 1989; Herman *et al.*, 1989; Selden and Song, 1994; Ayres and Ayres, 1996; De Bruyn *et al.*, 1997).

This study examines a system or network of five activities that are interrelated via material and product flows and market processes in a general equilibrium model. These activities may cause environmental damage by using resources and generating waste. Such impacts are represented as negative externalities that can be corrected for by taxes and subsidies. The model is unique in that recycling is considered as a separate decision-making activity, and that explicit material balance conditions are formulated. This is an extension relative to most previous studies on material and product flows, where either materials or products/markets are explicitly and completely described. As a result, the proposed model provides a general description of a material-product (M-P) chain on an aggregate level.

Ayres and Kneese (1969) studied materials in a general equilibrium model, but did not include products or taxation. A theoretical analysis of integrated resource extraction and pollution issues has been offered by Fullerton and Kinnaman (1995). They add a third option to the disposal options, garbage and recycling: namely, illegal dumping. Then, optimal taxation is based on a subsidy on garbage and a tax on the output of products. This may be interpreted as a general deposit-refund system. In this model a tax on new materials does not encourage recycling, but only optimizes the negative externality caused by using new materials. Most other studies on extraction and pollution focus on decisions made by households regarding recycling and dumping of garbage (Lusky, 1975 and 1976; Sullivan, 1987; Dinan, 1993; and Morris and Holthausen, 1994). In order to be able to provide a more complete description of the range of technological and allocative options, agents'

[1] This chapter is based on Kandelaars and Van den Bergh (1997b).

decisions regarding extraction, production, recycling of materials and products, and waste treatment need to be dealt with explicitly and with sufficient descriptive detail. In the analysis presented here, all agents' decisions associated with this range of interconnected processes are explicitly taken into account. Furthermore, material balance conditions are included, where relevant, to account for the physical properties of production and consumption. The goal of the analysis is to determine optimal tax rules in a market equilibrium. The framework of a general equilibrium model is regarded as useful because it clearly distinguishes between quantities, to which materials can be related, and prices, thus allowing a study of market-based policies.[2]

The type of analysis described here may focus on a specific material (e.g. a metal) or product (e.g. packaging). In order to reduce environmental pressure caused by pollution as well as to respond to an alarming scarcity of a natural resource, a strategy of reducing the net use of materials and products over time may be adopted. 'Net use' refers to gross materials use minus materials recycling and reuse of products, which should equal the materials obtained from resource extraction.

The remainder of this chapter is structured as follows. Section 7.2 presents the static general equilibrium model for a general M-P chain. Section 7.3 links the social welfare optimum to the market equilibrium so as to derive various optimal taxation rules. Conclusions are presented in Section 7.4.

7.2. A General M-P Chain

A general M-P chain is described in a static general equilibrium framework extended with material balance constraints. Decisions are made during five basic activities which are interrelated via material and product flows, and market processes. The five activities include extraction of resources, manufacturing of products, consumption of products, recycling and waste treatment. These activities are represented in the equilibrium model via separate profit and utility maximization formulations, and technical, budget and material balance restrictions. With this structure three important questions may be addressed: Via which price correction instruments or combination of such instruments (policy packages) can the external costs associated with resource extraction and waste disposal be optimized? What are the optimal price correction rules supporting the alternative policy packages? And, which second-best rules apply when specific instruments are not available or cannot be used?

The model presented is based on an M-P chain with a single product that is manufactured on the basis of a single new or recycled material. Once the product is economically depreciated, the consumer has two disposal options, namely recycling or dumping. The material that is recycled after use is assumed to be used as an input

---

[2] It is not necessary to believe that agents or sectors are accurately described by rational maximizing behaviour in a deterministic setting, but it is reasonable to expect that a general equilibrium framework, allowing for analytical solutions, could shed some 'qualitative light' on the relation between market-based environmental policy and material flows.

for the same product. A product itself cannot be reused. The income of consumers is spent on the main product generated by the M-P chain and on an alternative product. The latter may be interpreted as the total of all other products and services that are available for consumption, and whose material flows are not considered.

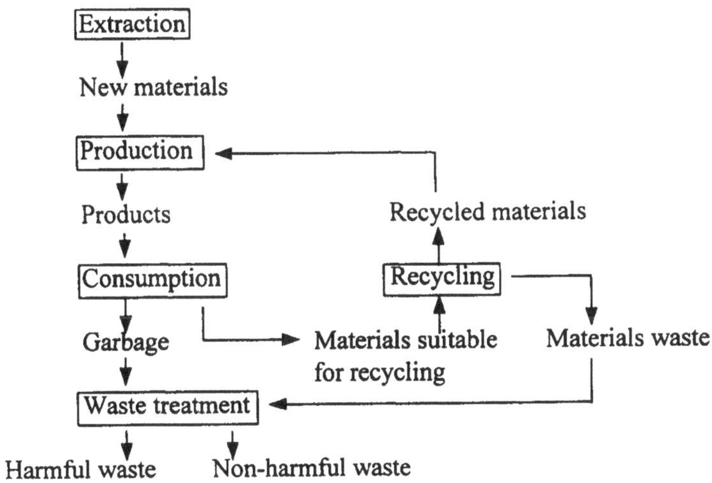

Figure 7.1. The material-product chain for one product.

The M-P chain has the following detailed structure. A main product is generated by a production process that uses three inputs: capital, new material and recycled material. The new material is obtained through the extraction of natural resources using capital. The alternative product only uses capital as an input. The two products are bought by n identical households. After use of the products in the consumption process these end up as garbage or as waste material entering a recycling process. This recycling process uses capital to recycle waste material, and generates waste material itself. The latter is treated in a waste treatment process together with the garbage directly originating from consumption. This leads to the distinction between harmful and non-harmful waste. The demand for the product by consumers results in two types of externalities: related to extraction and to harmful waste dumping. These externalities have a negative impact on the utility of the households: for example, cutting trees reduces the forest area and may reduce biodiversity, and waste dumping may cause health risks. In order not to complicate the model too much, it is assumed that the product itself cannot be reused, or that reuse is implicit in the market for the final product. Figure 7.1 presents the described M-P chain.

Each of the five types of activities, i.e. extraction, production of the final product, consumption of the final product, recycling of material and waste treatment are assumed to be characterized by behaviour consistent with the maximization of profits. All capital, material and product markets are assumed to be in equilibrium (markets clear and perfect competition applies). Furthermore, capital is assumed to be homogeneous among all activities. This allows for perfect capital mobility and emphasizes the long-run equilibrium character of the model. In order to be able to apply the material balance condition, all material variables (i.e. new and recycled material, garbage, material provided for recycling and harmful and non-harmful waste) have comparable material units.

The extraction process

The activity concerned with the extraction of the new material maximizes the net revenue of selling the new material minus the costs of capital used in the extraction process (with: V the amount of new material, p_V the price of new material, K_V the capital used for extraction, and p_K the price of capital).[3]

$$\max_{\{V,K_V\}} \; p_V V - p_K K_V \tag{1}$$

This maximization is constrained by the extraction process structure:

$$V = f^V(K_V) \tag{2}$$

with $f^V_{K_V} > 0$.[4] The first-order condition for a constrained optimum is:

$$p_V = \frac{p_K}{f^V_{K_V}} \tag{3}$$

The above formulation of the extraction process assumes that the amount of new material in the ground is unlimited or at least non-restrictive in the present model context. However, externalities associated with resource extraction, as well as the limited supply of capital, ensure that actual raw material use will be finite, both in the market and social optimum outcomes.

Production of the main product

The final product (X) is produced according to $f^X(K_X,V,R)$, with as inputs: productive capital (K_X), new material (V) and recycled material (R). The price of the product (p_X) is positive. The price of capital is p_K. The price of new material is p_V,

---

[3] The legend at the end of this chapter defines all variables, functions and parameters in the model. Variables that are measured in aggregated physical units (kilograms of materials) or functional units (functional units of products or units of capital) are represented by capital letters; prices and individual physical or functional units are presented in lower case letters. Functions describing process i are written as f^i.

[4] The notation is as follows. f^i_j is the first derivative of function f^i with respect to argument j, i.e. $f^i_j = \partial f^i(.)/\partial j$ for all i,j. It is assumed for all functions that the second derivative with respect to the inputs is non-positive (i.e. they are concave functions). Thus, function f^i holds, $\partial^2 f^i(.)/\partial j^2 \leq 0$, for all inputs j.

and in addition the use of this material is subjected to a tax rate t_V.[5] The use of recycled material is also taxed or subsidized (t_R) on top of its market price, p_R. In this model taxes (or subsidies) on new and recycled materials are imposed on the producers of the main product, because they choose between new and recycled materials in their production function.[6] It is assumed that all waste generated by the production of the final product is recycled directly within the production process.[7]

Profit maximization under perfect competition can be written as:

$$\max_{\{X,K_X,V,R\}} \; p_X X - p_K K_X - (p_V + t_V)V - (p_R + t_R)R \tag{4}$$

subject to the production function[8]

$$X = f^X (K_X, V, R) \tag{5}$$

where the first derivatives are $f^X_{K_X} > 0$, $f^X_V > 0$, and $f^X_R > 0$. The material inputs of the production function, V and R, may differ in quality implying that they are not fully substitutable.

The first order conditions lead to expressions involving the variables: price of capital, of new materials, of recycled materials and of the final product:

$$p_K = p_X \, f^X_{K_X} \tag{6}$$

$$p_V + t_V = p_X \, f^X_V \tag{7}$$

$$p_R + t_R = p_X \, f^X_R \tag{8}$$

Production of other products[9]

A second product is included to be interpreted as representing all the other products

---

[5] Note that if the tax rate $t_i < 0$, it may be interpreted as a subsidy.

[6] Another option would be to impose the tax on new materials on the extraction activity and the tax (subsidy) on recycled materials on the recycling activity.

[7] Without the assumption of direct internal recycling within the production process, i.e. no production waste, a multi-output production function is needed which allocates the material inputs among the final product and production waste. Then the material contents of the product will no longer be traceable, as a unit of product X does not include the total material inputs.

[8] This may be regarded as an extension of the formulation in Ayres and Kneese (1969) in which the material inputs of a product are fixed. More specifically, they use a (Leontief) production function with fixed coefficients, which implies that there is full complementarity between the inputs. The main disadvantage of this approach is that it does not allow for an analysis of substitution between materials and other inputs.

[9] An alternative product is needed because otherwise (1) consumers have to spend their whole income on the main product, and (2) the fixed capital stock is used within the M-P chain of the main product. Thus, the inclusion of the alternative product allows for substitution between the main and the alternative product and for a different allocation of the capital stock.

and services produced in the economy, which are not relevant from the perspective of the material flows and associated externalities focused on here. Additionally, it may be interpreted as also referring to use of leisure time and resources. A very simple linear production function is assumed for this alternative product. The alternative product only uses one input, namely capital. Every household consumes an amount a of the alternative product. Assuming n households, the total consumption is na.

$$na = K_A \tag{9}$$

Because of the linearity of the production function with one input, the price of product a equals the price of capital.

Households

There are n identical households whose utility function depends on the final product c, the aggregation of the other products in the economy a, the total amount of resource extraction V, and the total generation of harmful waste H. The latter two variables are included to account for the external costs caused by the related activities upon the households. This means that V and H cannot be influenced by individual households, and that the marginal utility of both V and H is negative ($u_V < 0$; $u_H < 0$). The welfare of households is positively affected by the consumption of products c and a ($u_c > 0$; $u_a > 0$). The maximization problem for each household is therefore:

$$\max_{\{c,a\}} u(c, a, V, H) \tag{10}$$

subject to a budget constraint and a material balance constraint. The budget constraint is as follows:

$$\frac{p_K\bar{K} + t_V V + t_R R + t_S S + t_G G + t_H H}{n} = p_X c + p_K a + (-p_S + t_S)s + (p_G + t_G)g \tag{11}$$

The left-hand side is the income of one household which includes capital earnings and transfers of tax revenues. Given the assumption of identical households, total income derived from capital ownership is equally distributed among households. The right-hand side includes the expenditures on the consumption of products c and a, and expenditures on, or revenues from, materials provided for recycling and garbage treatment. Households receive a payment p_S for the materials provided for the recycling process (explaining the negative sign of p_S). Garbage treatment and waste material for recycling are taxed or subsidized (t_G, t_S). The households choose to consume product c and divide their waste into garbage and recyclable waste (g and s). Their decisions depend on the prices of products c and a and on the costs or revenues resulting from g and s. Consumers may know their total waste material (s+g) because the producer gives that information to the consumer, or because the recycling activity can give that information to the consumer directly or via prices, assuming that the material contents of products is known by that activity.

The second constraint to the household decision problem is a material balance condition. It states that the material contents of product c ultimately end up in garbage (g), or in waste for recycling (s).[10] The output of materials of the consumption process is therefore s+g. The material inputs to the consumption process was V+R.[11] Of these material inputs V+R, (1/n)th part goes into product c.[12] Once new and recycled materials are part of a product, they cannot be distinguished any more. Therefore, for consumers only the material content of a product is relevant. The material balance condition is the following:[13]

$$s+g = \frac{V+R}{n} \tag{12}$$

After substitution of the material balance condition in (10) and (11), the Lagrange function associated with the household problem, given a shadow price of the budget constraint λ_1 and a shadow price λ_2 of the material balance condition, can be written as:[14]

Maximization with respect to c,a,s,g and λ_1 and λ_2 gives the following first-order conditions:

---

[10] A constraint that is related to the materials provided for recycling may be added. This constraint may reflect that the possibilities of recycling the waste material after use of the product depend on the amount of new and recycled materials that have gone into the production of the respective product. This can be formulated as $ns=f^s(V,R)$ with $f^s_V>0$ and $f^s_R<0$, reflecting that repeated recycling of materials leads to a lower quality of such materials. By adding this constraint to the model the amount of the waste material that will be provided for recycling is already determined by the function $f^s(V,R)$ and therefore prices and taxes cannot influence this division any more. Therefore, this constraint is not added in this model.

[11] The material inputs in the product are V and R, which are both measured in physical units (e.g. kilograms). These materials have a different quality for the producer (see eq. 5), but it is assumed that after consumption this quality difference is no longer relevant. Therefore, the materials content of a product equals (V+R)/n.

[12] The material contents of one product can be measured in two ways: (i) by the amount of inputs V+R and units X of the main product X: (V+R)/X; or, (ii) by the amount of consumption waste (s+g) and product c: (s+g)/c. Combining both of these with the market-clearing condition of final products, X=nc (see eq. 31) gives the material balance condition s+g=(V+R)/n (eq. 12).

[13] The material balance condition (eq. 12) implicitly includes two transformation functions: transformation of production materials into new products by means of the production function (f^X); and, the transformation of used products into waste material.

A special case is where new and recycled materials have the same quality. Then production equals $X=f^X(V+R,K_X)$ and the transformation from products to materials can be formulated as $nc=f^T(ns+ng,K_X)$. The function f^T can be seen as transforming the material contents of c that depends on the capital input in the production function into the waste materials s and g. It is immediately clear then that since the conditions (V+R)/n=s+g and nc=X (see eqs. 12 and 31) hold, the functions f^X and f^T are identical. This implies that either the MB condition or the market-clearing condition may be omitted. Since here a more general problem is studied, both of these conditions need to be included.

[14] The shadow prices λ_1 and λ_2 are not equal to 0, implying that the maximization problem is restricted by the budget condition and the material balance condition. The shadow price of the budget constraint λ_1 may be interpreted as the marginal utility of income.

$$\left.\begin{array}{c} L(c,a,s,g,\lambda_1,\lambda_2) = u(c,a,V,H) - \\ \lambda_1 \left[p_X c + p_K a + (-p_S+t_S)s + (p_G+t_G)g - \dfrac{p_K\bar{K}-t_V V-t_R R-t_S S-t_G G-t_H H}{n}\right] - \\ \lambda_2 \left[s + g - \dfrac{V}{n} - \dfrac{R}{n}\right] \end{array}\right\} \quad (13)$$

$$u_c - \lambda_1 [p_X] = 0 \quad (14)$$

$$u_a - \lambda_1 [p_K] = 0 \quad (15)$$

$$-\lambda_1 [-p_S + t_S] - \lambda_2 = 0 \quad (16)$$

$$-\lambda_1 [p_G + t_G] - \lambda_2 = 0 \quad (17)$$

and the budget constraint (11) with (12) substituted.

Recycling

Recycling uses two inputs, namely waste material provided for recycling, $S=ns$, and capital, K_R. The recycling function, f^R, has positive first derivatives, $f^R_S>0$ and $f^R_{K_R}>0$.

$$R = f^R(S,K_R) \quad (18)$$

A material balance condition associated with recycling ensures that the amount of direct waste of the recycling process, M, equals the input of materials, S, minus the level of recycled materials, R.

$$M = S-R \quad (19)$$

The recycling activity now maximizes the revenues of the recycled materials, $p_R R$, minus the costs of recycling, $p_S S$ and $P_K K_R$, and the costs of waste treatment, $p_M M$:

$$\max_{\{R,K_R,S,M\}} \; p_R R - p_K K_R - p_S S - p_M M \quad (20)$$

Substituting (18) and (19) in (20) results in:

$$\max_{\{K_R,S\}} \; (p_R+p_M)\, f^R(S,K_R) - p_K K_R - (p_S+p_M)S \quad (21)$$

The first-order conditions give:

$$(p_R+p_M)\, f^R_S = p_S+p_M \quad (22)$$

$$(p_R+p_M)\, f^R_{K_R} = p_K \quad (23)$$

Waste treatment

With G the waste of consumption ($G=ng$), the waste treatment activity maximizes the revenues of the total waste treated, $p_G G+p_M M$, minus the costs of capital use,

$p_K K_W$, and the costs of generating harmful waste.[15] These latter costs arise from a tax, t_H, being levied per unit of harmful waste generated, H. The optimization problem is formulated as:

$$\max_{\{G,M,K_w,H\}} \; p_G G + p_M M - p_K K_W - t_H H \tag{24}$$

subject to the waste treatment function and a material balance constraint. The waste treatment activity combines capital K_W, recycling waste M, and direct household waste G, as inputs thus realizing a certain level of harmful waste H. The first derivatives are $f^H_G > 0$, $f^H_M > 0$, and $f^H_{K_w} < 0$. This shows that the more capital is used, the less harmful waste will be generated.

$$H = f^H (G,M,K_W) \tag{25}$$

The material balance constraint states that final waste which is the sum of harmful waste H and non-harmful waste N, equals the sum of intermediate waste from consumption, G, and recycling, M.[16]

$$H+N = G+M \tag{26}$$

After substitution of the waste treatment function, the Lagrange function associated with the waste treatment problem, given a shadow price of the material balance condition μ (with $\mu <> 0$), can be written as:

$$\left.\begin{aligned} &L(G,M,K_W,\mu) = \\ &p_G G + p_M M - p_K K_W - t_H f^H(G,M,K_W) - \mu \, [N + f^H(G,M,K_W) - M - G] \end{aligned}\right\} \tag{27}$$

The first-order conditions are:

$$p_G = t_H f^H_G + \mu \, [f^H_G - 1] \tag{28}$$

$$p_M = t_H f^H_M + \mu \, [f^H_M - 1] \tag{29}$$

$$p_K = -t_H f^H_{K_w} - \mu \, [f^H_{K_w}] \tag{30}$$

and the material balance constraint (26).

Market clearing of final products

The n identical consumers each buy an amount c of the product making the total amount demanded nc. The production and supply of these final products is X. Therefore, the market-clearing condition is:

---

[15] Note that the revenues come from a part of the inputs (G and M), while the output (H) does not generate a direct revenue.

[16] Alternatively, N could have been chosen as the output of a treatment function, which is equivalent given (25) and (26) to $N = f^N(.) = M + G - f^H(.)$, so that $f^N_{K_w} > 0$.

$$X = nc \tag{31}$$

Market clearing of capital

The earlier mentioned homogeneous capital across activities and fixed supply, $\bar{K}$ together with an assumption of perfect capital markets leads to the following market-clearing condition:

$$\bar{K} = K_V + K_X + K_A + K_R + K_W \tag{32}$$

Material balance conditions

The material input in the production process is the sum of the new and recycled materials, V+R. Because it is assumed that the production process does not generate waste directly, the material contents of the products also equal V+R. Total material output of the consumption process is S+G. Together, these imply that the material balance condition associated with the subsystem of production and consumption activities can be written as V+R=S+G. Furthermore, the material balance condition associated with the recycling process is S=R+M (see eq. 19), and the one associated with the final waste treatment process is H+N=G+M (eq. 26). Substituting S and G from the latter two conditions in the first one, it follows that for the whole economy the following material balance condition results V+R=(R+M)+(N+H-M) or V=N+H, which means that the material input at the beginning of the M-P chain equals the output of materials at the end (see Figure 7.1). This straightforward result in a static setting confirms a consistent set of material balance conditions.

7.3. Optimal Tax Rules

In this section the market equilibrium (Appendix 7.A) and the social welfare optimum (Appendix 7.B) are compared to derive a set of optimal tax rules. For this reason, specifically equations B3 to B9 (Appendix 7.B) are combined with equations A2,A3,A5 and A7 to A11 (Appendix 7.A). Equations B10 to B12 of the social welfare optimum result from combining equations 5,9,12,26,31 and 32 of Section 7.2. These equations of Section 7.2 are also used in the market equilibrium (see Appendix 7.A). Therefore, these equations do not need to be combined because they are already equal.

First, the shadow prices δ_1, δ_2 and δ_3 are derived from equations B3,B5,B6,B9,A9 and A10.

$$\delta_1 = -\lambda_1 p_X \tag{33}$$

$$\delta_2 = nu_H - \lambda_1 p_X \frac{f^X_{K_x}}{f^H_{K_w}} \tag{34}$$

$$\delta_3 = nu_H + \lambda_1 p_X \frac{f^X_{K_x}}{f^H_{K_w}}(f^H_G - 1) \tag{35}$$

The optimal tax on the use of new material (t_V^*) is derived from equations A2,A3 and B7 and the shadow prices.

$$t_V^* = -\frac{nu_V}{\lambda_1} - \frac{nu_H}{\lambda_1} - p_X \frac{f^X_{K_x}}{f^H_{K_w}} (f^H_G - 1) \tag{36}$$

The external effects of the new material and harmful waste on utility are both negative ($u_V < 0$, $u_H < 0$), so that they contribute positively to the optimal tax. In other words, a larger external effect causes an increase in the optimal tax on the new material. The factor n incorporates the fact that n identical households suffer from the externality. The factor λ_1 translates utility to monetary terms. The third term in (36) is related to the second term in the optimal tax of the generation of harmful waste (t_H^*) in (37), derived from equation A5.

$$t_H^* = p_G + p_X \frac{f^X_{K_x}}{f^H_{K_w}} (f^H_G - 1) \tag{37}$$

The optimal tax t_H^* depends positively on the price p_G that is paid by the consumers for supplying waste to the waste treatment activity. An increase in p_G needs to be compensated by an increase in t_H^*.

The fact that the second term of t_H^* is identical to the third term of t_V^*, except for the sign, implies that a change in this term causes a shift between tax t_V^* at the start (i.e. extraction), and the tax t_H^* at the end (i.e. waste treatment) of the M-P chain. It is important to note that the use of the new material (V) is not equal to the generation of harmful waste (H) because a part of the output of the waste treatment function is non-harmful waste (N) (see the MB condition V=H+N derived in Section 7.2). This is reflected by the inclusion of f^H_G in the respective term.

The adjustment between t_V^* and t_H^* depends on the price of products and features of the production and waste treatment functions. The factor $f^X_{K_x}/f^H_{K_w}$ is the marginal productivity of capital in production relative to that in waste treatment. There are three possible effects of the second term of (37) on the taxes t_V^* and t_H^* that depend on the term f^H_G, that relates the garbage generated by the consumers to the harmful waste.

First, if $f^H_G < 1$, then for one unit increase of garbage less than one unit of harmful waste is generated (if everything else is fixed). If $f^H_G < 1$, then the contribution of the third term of (36) to t_V^* is negative and t_H^* is positive, because $p_X > 0$, $f^H_{K_x} > 0$ and $f^H_{K_w} < 0$. If f^H_G increases, t_V^* (t_H^*) will be lower (higher), because relatively more H will be generated. If $f^X_{K_x}/f^H_{K_w}$ decreases in absolute terms, then the optimal t_V (t_H) increases (decreases), because this change in the relative productivity causes less harmful waste to be generated and relatively more new material to be used. Finally, if the price of product (p_X) increases, the t_V^* (t_H^*)

will decrease (increase), because relatively more harmful waste is generated and less new material is used.

If $f^H_G>1$, an increase in garbage implies a more than proportional increase in harmful waste (note that this case is physically limited). The third term of (36) then contributes positively to t_V^* and the second term of (37) negatively to t_H^*. The effects are opposite to the case where $f^H_G<1$.

If $f^H_G=1$, which means that a change in G is proportional to a change in H, the t_V^* only depends on the external effects of V and H on the utility (u_V and u_H). The tax t_H^* equals p_G, because a change in H is proportional to a change in G (see also eq. 28 in Section 7.2). The generation of harmful waste (H) is related to the tax on using new material (t_V) by the proportional link between t_H and p_G in the waste treatment activity, the households (p_G, p_X) and production of the main product (p_X, p_V+t_V). Furthermore, any change in p_X or $f^X_{K_x}/f^H_{K_w}$ will not affect the trade-off between the two optimal taxes, due to the proportionality of changes in G and H.

Equation 37 combined with equations A8 and B8, and the shadow prices, results in an expression for the optimal tax on using recycled material (t_R^*):

$$t_R^* = p_G \tag{38}$$

In the optimum t_R^*, equals the price for the generation of garbage paid by the consumers. With a policy on recycled material t_R^*, the price p_G switches the tax to the end of the chain.

When $p_G=t_R^*>0$, the combination of t_R^* and p_G in the optimum may be interpreted as a deposit that the producers pay for using recycled material (t_R^*) that will be refunded to the waste treatment activity once waste material is collected for garbage (p_G). If p_G decreases, consumers will provide less waste material to recycling, so that the supply of recycled material will decrease, and the optimal tax on recycled material will decrease (t_R^*). When p_G is negative, i.e. a revenue for the households and a cost for the waste treatment activity, then t_R^* is a subsidy.

Combining equations B4,A5,A7,A11 and the shadow prices, a relationship for the taxes on garbage (t_G^*) and material provided for recycling (t_S^*) is derived in (39). These two taxes cannot be chosen independently. The costs or revenues for households of dividing their waste material after consumption between S and G are equal with taxation (see eq. A11). If these were not equal, then households would decide to either recycle or dump all waste material. Therefore, the taxes t_S^* and t_G^* are directly related. In fact, there is an unlimited set of combinations of taxes. If t_G^* increases, t_S^* will increase by the same amount to maintain the optimal choices by consumers. The taxes depend on features of the production and recycling functions, and also on the price of the product.

$$t_S^* - t_G^* = p_X \frac{f^X_{K_x} f^R_S}{f^R_{K_R}} - p_X \frac{f^X_R f^R_S}{1-f^R_S} \quad \text{for } f^R_S \neq 1 \tag{39}$$

The first term in (39) is positive because all elements are positive. An increase in the marginal productivity of capital in the production function relative to that marginal productivity in the recycling function indicates that it is better to recycle less material. Then, $t_S^*-t_G^*$ will increase so that less material will be provided to the recycling activity and thus more material will go to the waste treatment activity.

The sign of the second term of (39) depends on f^R_S, which is the marginal productivity of waste material in the recycling function, and has three possible effects on $t_S^*-t_G^*$: negative if $f^R_S<1$; positive if $f^R_S>1$; and, a special case if $f^R_S=1$.

First, the expected case of $f^R_S<1$ is described, which means that for one unit of waste material less than one unit of recycled material is produced. An increase in the f^R_S, meaning that less waste material is needed to produce one unit of recycled material, affects the optimal taxes $t_S^*-t_G^*$ depending on $f^X_{K_X}/f^R_{K_R}$, the relative marginal productivity of capital in the production and recycling functions, and on f^X_R, the marginal productivity of recycled material in the production function. The term $f^X_R f^R_S/(1-f^R_S)$ may be interpreted as the impact of the production function on the waste material for recycling (S) via repeated recycling. This may be seen by rewriting $f^X_R f^R_S/(1-f^R_S) = f^X_R(f^R_S+f^{R\,2}_S+f^{R\,3}_S+f^{R\,4}_S+\ldots)$. An increase in f^R_S causes an increase in $f^R_S/(1-f^R_S)$. The total effect on t_S-t_G may be positive, negative or zero depending on the exact value of each marginal productivity in (39). For an increase in p_X the effect on the optimal $t_S^*-t_G^*$ depends on the marginal productivities and may also be positive, negative or neutral.

When $f^R_S>1$, which means that an increase in S gives a more than proportional increase in R, $t_S^*-t_G^*$ is always positive and thus $t_S^*>t_G^*$. Note that this case is limited by physical constraints. If f^R_S increases, then for the same amount of R less S is needed and therefore $t_S^*-t_G^*$ increases. The effects of changes in the second term have an opposite effect on $t_S^*-t_G^*$ than if $f^R_S<1$.

If $f^R_S=1$, for one unit of S one unit of R is produced. In this case equation 39 is not valid. Equation B4 gives that $\lambda_1 p_X f^X_R=0$, so then it follows that $\lambda_1=0$, because p_X and f^X_R are positive. This may be interpreted such that the MB condition is stricter than the budget constraint which is not a realistic case (see eq. 13).[17]

Second-best taxes

It may be impossible to impose optimal (first-best) taxes. One reason is that a tax on new material cannot be imposed because a part of the M-P chain is abroad, or because a certain tax is politically not acceptable. Then, second-best taxes may be imposed. To obtain the second-best taxes, equations 36 to 38 are combined. This gives the following relationship of suboptimal taxes denoted as $t_i^{\#}$ for a tax on i.

$$t_V^{\#} + t_H^{\#} - t_R^{\#} = -\frac{nu_V}{\lambda_1} - \frac{nu_H}{\lambda_1} \tag{40}$$

This shows the possible trade-off or compensation of taxes in a second-best world, i.e. if one or more of the taxes cannot take its optimal value as indicated by equations 36 to 38.

---

[17] It has also been examined what would happen if a tax on products were included. In equations 11,13 and 14 (see Section 2) the term p_X will then be replaced by p_X+t_X. Thus, equation A9' will become $u_c-\lambda_1[p_X+t_X]=0$. Comparing equation A9' with B3 using the shadow price δ_1 (equation 33), the optimal level of t_X is zero, i.e. this tax is independent of prices or other taxes. In the optimal situation, the externalities caused by the M-P chain are covered by other taxes. A non-zero value of t_X would lead away from the social optimum.

If $t_V^{\#}$ cannot be imposed, because extraction happens abroad and a tax on the border is in contravention of trade agreements, then the externalities are suboptimized by imposing $t_H^{\#}$ and $t_R^{\#}$. It may be impossible to impose an optimal $t_H^{\#}$ because of imperfect monitoring of harmful waste, so that the likelihood of illegal dumping arises. If $t_H^{\#}$ cannot be imposed $t_V^{\#}$ will increase or $t_R^{\#}$ will decrease compared with the first-best solution. If $t_R^{\#}$ cannot be imposed, then $t_V^{\#}$ and $t_H^{\#}$ will change compared with the first-best solution. If $t_R^{\#}$ was negative (i.e. a subsidy) in the first-best solution, then $t_V^{\#}$ and $t_H^{\#}$ will increase compared to that optimum. Lastly, if two of the three taxes cannot be imposed, one may end up further away from the social optimum.

7.4. Conclusions

The approach here is new in the sense that it bridges the literature on material flow analysis and environmental policy analysis based on externality/welfare theory. Although some studies have already combined some elements of both these approaches, the combination of extraction, production, recycling, consumption, waste treatment activities and material balance conditions in a general equilibrium framework, as pursued in this chapter, is new. It was shown to permit the derivation of the optimal taxation and subsidization rules which take account of flows and processes related to the new material, recycled material, main product, garbage from consumption, and material and recycling waste (see Figure 7.1).

The main conclusions are as follows. The externalities caused by the use of a new material and the generation of harmful waste are optimized by taxing a new material. Hence, the optimum does not allow for a tax on other materials, product or capital inputs or output of the M-P chain to optimize the externalities. In a second-best world the externalities may be suboptimized by taxing the generation of harmful waste or by subsidizing the use of recycled material. The optimal tax on the generation of harmful waste and the tax on a new material depend partly on the same term that includes prices and marginal products of the production and waste treatment functions, except for the sign of this term. This implies that a change in one of the factors causes a shift between the optimal taxes on a new material at the beginning and on harmful waste at the end of the M-P chain. This linkage is interesting because it shows that the whole M-P chain needs to be considered instead of parts of this chain. The optimal taxes on garbage and material intended for recycling cannot be chosen independently, because the decision to divide the waste material into garbage and material intended for recycling depends directly on both taxes. The optimal taxes depend on marginal products of the production, recycling and waste treatment functions. The optimal tax on recycled material equals the price consumers pay for garbage collection. Thus, this tax translates the end (here, the waste generation) of the M-P chain to the beginning (here, the use of recycled material).

In this chapter the focus has been on deriving general tax rules in a market model, with an average type of household and general functional specifications. In addition, comparative static analysis, based on specific functional specifications, can be pursued to assess the effects of changes in consumer preferences and technological options in extraction, production, recycling and waste treatment.

In order to create a more realistic consumer behaviour setting, imperfect products may be added, to allow the consumer to choose between (imperfectly substitutable) products which satisfy similar needs, but are essentially different in terms of their material composition. Another extension is considering interaction between different types of materials, by modelling products composed of multiple materials. This may also allow for trade-offs between externalities associated with different materials, and addresses the problem of shifting between materials and associated environmental problems. Finally, the present approach deals only with static externalities. It is also interesting to consider dynamic externality effects such as delayed material flows in product life cycles, vintages of a product with a different material composition and the accumulation of materials in the economy. This may be based partly on analytical models and partly on numerical simulation under different future scenarios about technological progress concerning processes and products and demand-side developments.

Legend

	Name	Unit[18]	Comments	Number of variables, First derivatives
Volumes	V	Kilograms	New material	1
	R	Kilograms	Recycled material	1
	K_i	Capital unit	Demand for capital of i	5 (index i: V,R,X,A,W)
	X	Functional unit	Supply/Production of final product	1
	c	Functional unit	Individual household demand for product	1
	a	Functional unit	Individual household demand for alternative product	1
	S (s)	Kilograms	Aggregated (individual) waste provided for recycling	2
	M	Kilograms	Recycling waste	1
	G (g)	Kilograms	Aggregated (individual) garbage	2
	N	Kilograms	Non-harmful final waste	1
	H	Kilograms	Harmful final waste	1
Prices	p_i	Monetary	Price of i	6 (index i: V,R,K,S,G,M)
	p_X		Price of product ($p_X>0$)	1
	λ_i,μ		Shadow prices	3 (index i: 1,2)
Functions	u		Utility function of an individual household	$u_c,u_a>0$; $u_V,u_H<0$
	f^V		Extraction function	$f^V_{K_V}>0$
	f^X		Production function	$f^X_{K_X}, f^X_V, f^X_R>0$
	f^R		Recycling function	$f^R_S, f^R_{K_R}>0$
	f^H		Waste treatment function	$f^H_G, f^H_M>0$; $f^H_{K_W}<0$
Parameters	t_i	Monetary	Tax (subsidy) on i if $t_i>0$ (<0)	6 (index i: V,R,H,C,G,S)
	n	Integer	Number of households	
	$\bar{K}$	Capital unit	Fixed supply of capital	

[18] All kilograms, capital units, functional units are non-negative.

Appendix 7.A. The Market Equilibrium

The market equilibrium with taxes and subsidies is the solution of the following set of equations together with equations 9, 11, 12, 26, 31 and 32 (see Section 7.2).
From equation 6 of Section 7.2, it follows:

$$p_K = p_X f^X_{K_x} \tag{A1}$$

From equations 3 and A1, it follows:

$$p_V = p_X \frac{f^X_{K_x}}{f^V_{K_v}} \tag{A2}$$

From equations 7 and A1, it follows:

$$p_V = p_X f^X_V - t_V \tag{A3}$$

From equations 8 and A1, it follows:

$$p_R = p_X f^X_R - t_R \tag{A4}$$

From equations 28-30 and A1, it follows:

$$p_G = - p_X \frac{f^X_{K_x}}{f^H_{K_w}} (f^H_G - 1) + t_H \tag{A5}$$

$$p_M = - p_X \frac{f^X_{K_x}}{f^H_{K_w}} (f^H_M - 1) + t_H \tag{A6}$$

From equations A1,A6,22 and 23, it follows:

$$p_S = p_X \frac{f^X_{K_x}}{f^R_{K_x}} f^R_S + p_X \frac{f^X_{K_x}}{f^H_{K_w}} (f^H_M - 1) - t_H \tag{A7}$$

From equations A1,A4,A6 and 23, it follows:

$$p_X f^X_R - t_R = p_X \frac{f^X_{K_x}}{f^R_{K_x}} + p_X \frac{f^X_{K_x}}{f^H_{K_w}} (f^H_M - 1) - t_H \tag{A8}$$

From equations 14-17 and A1, it follows:

$$u_c - \lambda_1 [p_X] = 0 \tag{A9}$$

$$u_a - \lambda_1 [p_X f^X_{K_x}] = 0 \tag{A10}$$

$$-p_S + t_S = p_G + t_G \tag{A11}$$

Appendix 7.B. The social welfare optimum

In the social welfare optimum the utility of the n identical households is maximized (assuming identical households and therefore applying equal weighting):

$$\max_{\{c,a,V,H\}} \; nu\,(c,a,V,H) \tag{B1}$$

subject to the production functions, transformation function and material balance conditions of Section 7.2. Contrary to the market equilibrium, in the social welfare optimum the externalities (V and H) are control variables. A Lagrange function is defined based on shadow prices $(\delta_1,\delta_2,\delta_3)$ for the market clearing condition (X=nc; eq. 31), the MB condition of final waste (N+H=M+G; eq. 26) and the MB condition for consumption (S+G=V+R; eq. 12), respectively. After substituting the other constraints (eqs. 2,5,9,18,19 25 and 32 of Section 7.2), the Lagrange function can be written as:

$$\left.\begin{array}{c} L(c,a,s,g,K_V,K_R,K_W,\delta_1,\delta_2,\delta_3) = \\ n\;\underline{u}[c,a,f^V(K_V),f^H(ng,ns-f^R(ns,K_R),K_W)] - \\ \delta_1\;[f^X[\bar{K}-K_V-na-K_R-K_W,f^V(K_V),f^R(ns,K_R)]-nc] - \\ \delta_2[N+f^H(ng,ns-f^R(ns,K_R),K_W)-ns+f^R(ns,K_R)-ng] - \\ \delta_3\;[ns+ng-f^V(K_V)-f^R(ns,K_R)] \end{array}\right\} \tag{B2}$$

The first-order conditions are:

$$nu_c + \delta_1 n = 0 \tag{B3}$$

$$\left.\begin{array}{c} nu_H f^H{}_M(n-f^R{}_S n) - \delta_1\;[f^X{}_R f^R{}_S n] - \\ \delta_2\;[f^H{}_M(n-nf^R{}_S)-n+f^R{}_S n] - \delta_3\;[n-f^R{}_S n] = 0 \end{array}\right\} \tag{B4}$$

$$nu_H f^H{}_G n - \delta_2\;[f^H{}_G n-n] - \delta_3[n] = 0 \tag{B5}$$

$$nu_a + \delta_1\;[f^X{}_{K_x} n] = 0 \tag{B6}$$

$$nu_V f^V{}_{K_v} - \delta_1\;[-f^X{}_{K_x}+f^X{}_V f^V{}_{K_v}] - \delta_3\;[-f^V{}_{K_v}] = 0 \tag{B7}$$

$$\left.\begin{array}{c} nu_H f^H{}_M(-f^R{}_{K_R}) - \delta_1\;[-f^X{}_{K_x} + f^X{}_R f^R{}_{K_R}] - \\ \delta_2\;[-f^H{}_M f^R{}_{K_R} + f^R{}_{K_R}] - \delta_3\;[-f^R{}_{K_R}] = 0 \end{array}\right\} \tag{B8}$$

$$nu_H f^H{}_{K_w} + \delta_1\;[f^X{}_{K_x}] - \delta_2\;[f^H{}_{K_w}] = 0 \tag{B9}$$

$$f^X[\bar{K}-K_V-na-K_R-K_W,f^V(K_V),f^R(ns,K_R)] = nf^c(s+g) \tag{B10}$$

$$N + f^H(ng,ns-f^R(ns,K_R),K_W) = ns-f^R(ns,K_R)+ng \tag{B11}$$

$$ns + ng - f^V(K_V) - f^R(ns,K_R) = 0 \tag{B12}$$

CHAPTER 8

A DYNAMIC ANALYSIS OF RAIN GUTTERS[1]

8.1. Introduction

In this chapter a dynamic model is presented in order to study the impacts on M-P chains of various economic and policy developments aimed at reducing material use. In a dynamic analysis temporal issues can be taken into account. Dynamic processes that may be important in an M-P analysis are accumulation and delays. Accumulation of products and materials may occur in the economy: for example, for durable products there is a time-lag between the purchase and the disposal of the product. Delays may also occur between the invention and the adoption of that technology, and between the implementation and the effect of a policy. The dynamic model aims at tracing the effects of changes in material flows over time: for example, through substitution and recycling. It is used to simulate policy scenarios so as to assess their short- and long-term influence on the M-P chain. All these accumulative and delayed effects may not be adequately analysed in a static setting. The dynamic model will be applied to rain gutters.

Dynamics related to material and product flows also feature in an alternative, Neo-Austrian economic theory, as shown in Faber *et al.* (1987). These authors provide accurate descriptions of the dynamics of production, reuse of products and recycling of materials, based on a combination of economic and thermodynamic insights, using both material balance and entropy process formalizations. In a somewhat similar approach, Van den Bergh and Nijkamp (1994) link the dynamic development of an economy - in terms of man-made capital used for extraction, capital production, goods production, recycling of materials, and technological-knowledge formation - to production functions with material and non-material inputs and dynamic material balance conditions. These approaches are especially useful to investigate the relationship between resource use and pollution problems in the long run via material flows.

Another type of dynamic analysis is an econometric analysis based on historical data. Kandelaars and Van Dam (1999) analyse the use of materials in automobiles. The statistical analysis shows that the price of materials does not influence its use. This implies that a material policy to reduce the use of specific materials is probably not effective.

In Section 8.2 the M-P chain for rain gutters is explained. In Section 8.3 the time pattern for gutters is formally described. A dynamic simulation model for rain

[1] This chapter is a slightly revised version of an article that appeared in *Journal of Environmental Systems* (Kandelaars, Opschoor and Van den Bergh, 1996).

gutters is presented in Section 8.4. In Section 8.5 the scenarios, the control variables and the indicators of the model are discussed. Section 8.6 discusses the results of the scenarios of the analysis. A final section offers conclusions.

8.2. Rain Gutters as a Case-Study

This chapter focuses on one particular class of products, i.e. gutters, which provide one particular service, i.e. removing water from the roofs of houses. In this M-P chain two types of rain gutter will be compared, i.e. zinc (a metal) and polyvinyl chloride (pvc) gutters. The case of gutters is chosen because zinc gutters consume a considerable part of the total amount of zinc used in the Netherlands. Until recently zinc was not considered to cause serious environmental problems; this opinion is somewhat related to the fact that it is a trace element needed by human beings, animals and plants. However, two main problems have emerged as a result of the increased use of zinc: (1) high concentrations of zinc may harm crops in agricultural soil and organisms in surface water; and (2) the stock of zinc ore in the environment is exhaustible (Gorter, 1994). The main environmental problems caused by zinc are listed as aquatic toxicity, acidification, landscape deterioration and the extraction of the toxic material cadmium, which is a trace element in zinc ore (Tauw Milieu, 1994 and Jolly, 1992).

The most important environmental impacts of the alternative type of gutter which is made of pvc, are the use of energy and the releases of several toxic substances, such as chlorine. Moreover, the transport of chlorine by rail causes a serious environmental risk (Heijnis, 1990). In this chapter, however, it is assumed that reducing the use of zinc has a higher priority than reducing the use of pvc. Therefore, the central focus of this chapter is to reduce the use of zinc. Although both pvc and zinc gutters have a negative impact on the environment, the choice between zinc and pvc gutters is mainly determined on the basis of their prices and the preferences of the consumers, in particular construction firms.

In the Netherlands the market for gutters is dominated by zinc gutters because they are strong, attractive, cheap, and used traditionally. In several other countries of the European Union this dominating position is not found: for instance, in France pvc gutters dominate the market (Fraanje and Verkuijlen, 1996).

A 'gutter service' will be measured in functional units. The functional unit chosen is the length of gutter needed for an average house, which is 12 metres (Fraanje and Verkuijlen, 1996). The weight of this functional unit depends on the type of gutter: a zinc gutter of 12 metres weighs 29.6 kilograms and a pvc gutter of the same size 16.1 kilograms (Fraanje and Verkuijlen, 1996).

A gutter service requires three elements: a gutter, a fastening-piece and a waste-pipe. Besides the gutters, the fastening-pieces are of substantial interest, because the amount of material needed for a fastening-piece depends on the type of gutter. The fastening-pieces for both types of gutter are made of the same material, i.e. galvanized steel. However, these fastening-pieces differ in weight because zinc gutters are heavier than pvc gutters and therefore need stronger and hence heavier fastening-pieces. The waste-pipe is not included in the model because such pipes are the same for both types of gutter and therefore irrelevant to a comparison.

Zinc gutters are not reused as such, but a large part of the zinc content of a gutter is recycled. Zinc gutters consist of 99.8% zinc, which means that recycling through melting can be done without serious problems. In the Netherlands used zinc gutters are collected and transported to Germany for recycling. The zinc which is imported in the Netherlands is partly made of zinc ore and partly of remelted zinc. However, no data are available on the proportion of the zinc which is made from remelted zinc, i.e. recycled or secondary zinc (Gorter, 1994).

In addition to zinc gutters, pvc gutters are used to meet the demand for gutters. The differences between pvc and zinc gutters are that (1) pvc gutters have a shorter life-span than zinc gutters and (2) the quality of pvc gutters is inferior to that of zinc gutters. After use, pvc can be recycled provided that it is not polluted with other materials. The presence of other materials changes the material characteristics which limits the recycling possibilities. In the case of pvc gutters the pvc is polluted with glue, which makes high-grade recycling impossible. As opposed to the recycling of zinc, the recycling of pvc is not profitable given current prices, i.e. the costs of producing new pvc are lower than those of recycling pvc.

The fastening-pieces of both zinc and pvc gutters are made of galvanized steel which is as yet not recycled for technical and economic reasons. Research is being done, e.g. at Hoogovens, to make it technically possible and economically viable to recycle the layer of zinc from the galvanized steel (Fraanje and Verkuijlen, 1996). The quantities of zinc, pvc and galvanized steel that are not recycled enter the environment as waste flows.

Data on the production, use and disposal of gutters for the base year can be found in Fraanje and Verkuijlen (1996), Tauw Milieu (1994), Matthijssen and Meijer (1992) and Gorter (1994). The base year of this study is 1990 because it is the most recent year for which a complete set of data is available. The time horizon of the model simulation is set at 60 years, in order to allow for a comparison of the two types of gutter that have specific lifetimes: 30 years for zinc gutters and 20 years for pvc gutters.

8.3. The Use of Rain Gutters over Time

In this section a formal description is given of the demand for gutters and the stock of gutters in use over a period of time. The demand for gutters depends on the number of replaced gutters and the number of newly built houses for which gutters are needed.

The yearly demand for gutters, D_t, consists of replaced gutters, R_t, and gutters needed for newly built houses, N_t.

$$D_t = R_t + N_t \qquad (1)$$

The number of gutters that is used up at time t is equal to the demand at (t-T), D_{t-T} with T equal to the lifetime of gutters. D_{t-T} by definition originates from existing houses. These can be renovated which involves a replacement of the gutters. This replacement demand is R_t. Alternatively, these houses can be demolished rather than renovated, which gives a flow of disposed gutters, W_t. Thus,

$$R_t+W_t=D_{t-T} \tag{2}$$

The replacement demand, R_t, is a fraction s_t of the amount of gutters that is used up:

$$R_t=s_tD_{t-T} \tag{3}$$

The total stock of gutters at time t, G_t, is equal to the stock of gutters at time (t-1), G_{t-1}, plus the new gutters at time t, N_t, minus the disposed gutters at time t, W_t.

$$G_t=G_{t-1}+N_t-W_t \tag{4}$$

From equations (1) and (2) it follows that $N_t-W_t=D_t-D_{t-T}$. With (4) this results in:

$$G_t=G_{t-1}+D_t-D_{t-T} \tag{5}$$

With $D_t=0$ for $t\leq 0$, repeated substitution of (5) leads to:

$$G_t=\sum_{\tau=t-T+1}^{t} D_\tau \tag{6}$$

The total stock of gutters, G_t, is equal to the demand for gutters of the last T years, since T is the lifetime of gutters.

Figure 8.1 shows time patterns for the most relevant variables, and relates them to one another. These variables are: (1) the gutters needed for newly built houses at time t, N_t; (2) the replaced gutters at time t, R_t; (3) the total demand at time t, D_t; and, (4) the stock of gutters at time t, G_t.

In the simulation model zinc gutters are assumed to have a lifetime of 30 years, which implies that the demand for zinc gutters in 1960 is replaced 30 years later, i.e. in 1990. The time horizon of the model is 60 years. Therefore, the demand for and the stock of zinc gutters is needed for the period 1960-2050; the years 1960-1989 are used to calculate the number of zinc gutters that need to be replaced and the years 1990-2050 to obtain the number of zinc gutters that are needed. For pvc gutters the lifetime is assumed to be 20 years.

8.4. Model Description

Demand for gutters is assumed to be given and met by two substitutes: zinc and pvc gutters. These two types of gutter form the basis of the model which consists of eight submodels: three at the product level, three at the material level, one for the extraction of ores and one for the calculation of prices and costs. Figure 8.2 shows these eight submodels and their links. Appendix 8 contains details about the initial conditions and specific equations in the model.

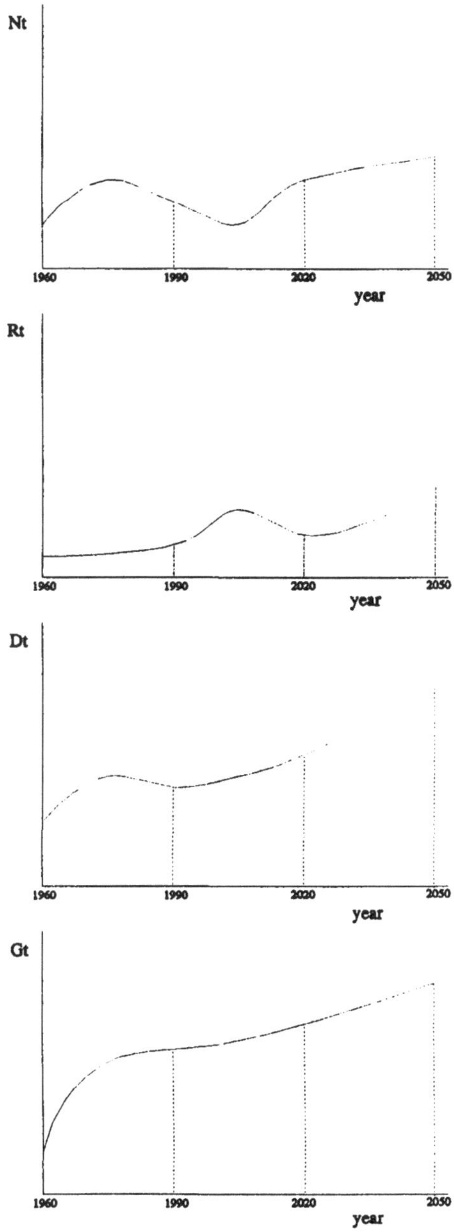

Figure 8.1. The demand for and the stock of gutters at time t.
Note: Variables are explained in the text (Section 8.3).

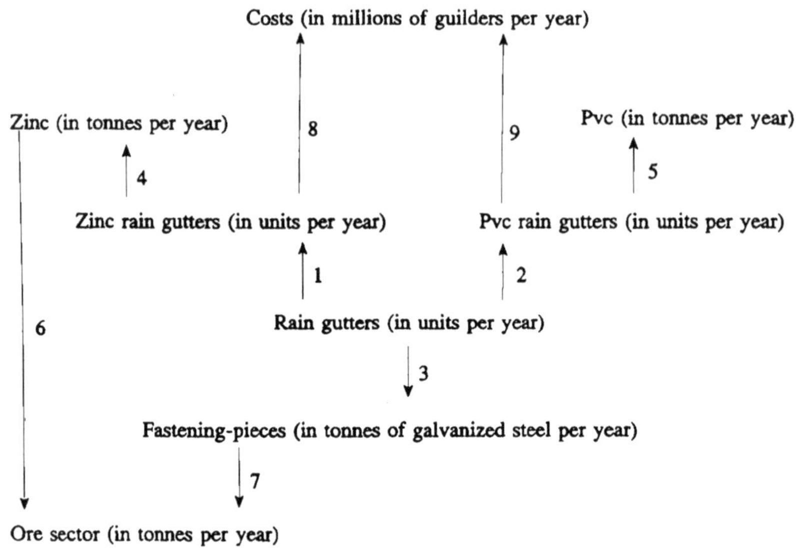

Figure 8.2. Basic simulation model for linking the demand for rain gutters to material flows.
Note: Numbers are explained in the text (Section 8.4).

Demand for rain gutters

In this part of the model the demand, the accumulation in the economy and the waste of gutters are modelled. In the product submodels, gutters are measured in functional units. In the model the yearly total demand for gutter services which has to be met is given exogenously. In the base year 1990 the number of gutters in the economy is 2,660,560. Every year some of these gutters are replaced and some are demolished, because the house to which they are affixed is destroyed. The replaced gutters and the gutters needed for newly built houses form the yearly demand for gutters. Note that gutters are not yet recycled at a product level.

Zinc and pvc rain gutters

These two submodels describe the production of zinc and pvc gutters. The demand for gutters is allocated to zinc and pvc gutters (arrows 1 and 2 in Figure 8.2). This division of the demand for gutters into the two types of gutter is one of the major decision variables of the model. This division variable also determines the quantities of zinc and pvc needed to satisfy the demand for gutters. The total demand for gutters needs to be met by zinc and pvc gutters.

Fastening-pieces

A complete rain gutter service consists of fastening-pieces which are made from galvanized steel. The fastening-pieces are measured in tonnes of galvanized steel for both types of gutters. The number and types of fastening-pieces which are made, demolished and renovated depend directly on the allocation of the demand for gutters (arrow 3 in Figure 8.2). As opposed to pvc and zinc, galvanized steel is not recycled.

Zinc, pvc and galvanized steel

Here the quantities of zinc, pvc and galvanized steel are measured in tonnes. The zinc (pvc) submodel is directly linked with the zinc gutter (pvc gutter) submodel, indicated by arrow 4 (5) in Figure 8.2. The gutters (measured in functional units - see Section 8.2) are connected with the submodels for zinc, pvc and galvanized steel by converting the number of products into the amount of materials. The conversion from functional units to tonnes is done by a conversion factor. The materials needed for both types of gutter are calculated at the product level. Figure 8.3 illustrates the flows of zinc from the environment through the economy and back to the environment.

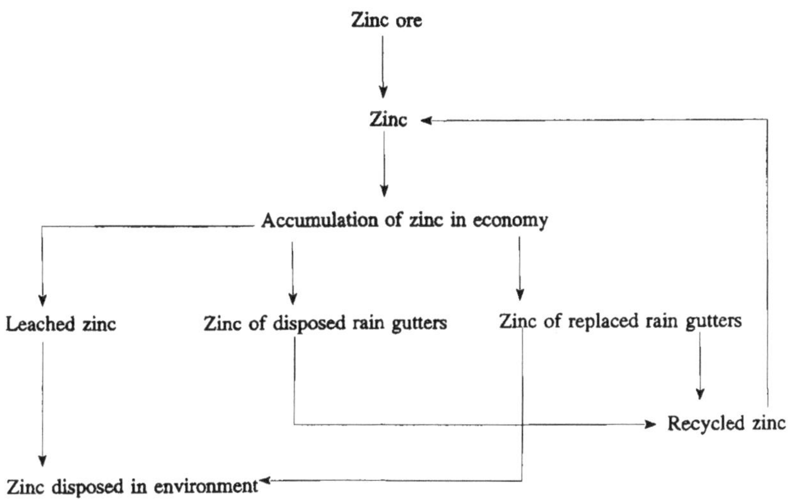

Figure 8.3. A material flow model of zinc in the environment and the economy.

Ore sector

A submodel for metal ores is added to describe and analyse the rate of extraction needed to meet the demand for gutters. Extraction may result in the depletion of zinc ore. Ores are needed for the production of zinc and of galvanized steel (arrows 6 and 7 in Figure 8.2), which means that the extraction rate depends on the demand for zinc and galvanized steel. Ores are measured in tonnes.

Prices and costs

In this economic submodel the costs of meeting the demand for the M-P chain and the revenues of reusing materials are calculated. The costs are directly linked to the use of zinc and pvc gutters (arrows 8 and 9 in Figure 8.2). The costs involved are the prices of the two types of gutter and the revenues of used materials (zinc, pvc and galvanized steel). These revenues together with the prices of zinc and pvc gutters determine the allocation of the demand between the two types of gutter. The net costs of a gutter equal the costs of buying a gutter minus the revenues of used materials. The price of one kilogram of new zinc equals the price of recycled zinc, which makes it reasonable to assume that the difference between the price of recycled zinc and the revenue of used zinc equals the costs of recycling.

8.5. Scenarios, Control Variables and Indicators

Five scenarios are analysed by way of dynamic simulation. These scenarios focus on the influences of preferential policy and economic change on material flows, through changes in M-P chains.

The following scenarios are studied:

1. Base scenario;
2. Preferences shift scenario;
3. Product charge scenario;
4. Recycling scenario; and,
5. Mixed scenario.

The base scenario, in which the exogenously determined variables are held constant, serves as a reference scenario for the other scenarios. The control variables are: (1) the allocation variable that distributes the demand for gutters between the two types of gutter; (2) the price of a zinc gutter that influences the costs of meeting the demand and the distribution of the demand; (3) the price of recycled zinc that influences the cost and the percentage and quantity of zinc that is recycled; and, (4) the price of recycled pvc that influences the cost and the percentage and quantity of pvc that is recycled.

Table 8.1 summarizes the scenarios and the associated levels of the control variables. The second column shows the values of the control variables in the base scenario; for the other four scenarios the values of the control variables are only given when they differ from the base scenario.

For a comparison of the various scenarios the following clusters of performance indicators are distinguished: (1) the quantities extracted from and disposed of into the environment; (2) the levels of recycling; (3) the allocation of the demand for gutters; and, (4) the prices and the net costs of satisfying the gutter demand.

Base scenario

In the base scenario all exogenous variables remain stable over time at the level of the base year 1990: demand for gutters, allocation over zinc and pvc gutters, recycling percentages for pvc, zinc and galvanized steel, demolition and replacement of gutters. The data for the base scenario are obtained from various sources mentioned in the model description in Section 8.2.

Table 8.1. The scenarios with their control variables.

Control variable \ Scenario	Base	Preferences shift	Product charge	Recycling	Mixed
Allocation variable	0.8	max (0.8-0.03t,0.2)			
Price of zinc gutter	180		min (180+5t,300)		min (180+5t,300)
Price of recycled zinc	1			2	2
Price of recycled pvc	0			1	1

Preferences shift scenario
In the preferences shift scenario the allocation variable is set at a different level than in the base scenario. The allocation variable is assumed to change over time in order to be able to analyse the impact of a change in the distribution of the demand on economic and environmental variables. Therefore, the distribution of the demand for gutters between zinc and pvc gutters changes over time. In this scenario the total demand for gutters is met increasingly by pvc gutters, which could be the result of more information or education by the government on the environmental implications of the two types of gutter. In this scenario the percentage of the demand for zinc gutters changes from 80 to 20% in 20 years, i.e. max (0.8-0.03t,0.2) in Table 8.1, while all the other variables stay the same, e.g. the price for a zinc or pvc gutter remains Dfl 180. The shift in the allocation of the demand alters the quantities of zinc and pvc needed and in turn the extraction from and disposal into the environment. The percentages of recycling do not change.

Product charge scenario
This scenario differs from the base and preferences shift scenarios in that the allocation of the demand depends on the prices of the two types of gutter. In this scenario the price of zinc gutters is raised because of a charge which increases over time up to a certain limit, i.e. min(180+5t, 300) in Table 8.1. This charge influences the distribution of the demand over the two types of gutter. In reality the world price of zinc has been decreasing. However, in this scenario the government is assumed to raise the price of zinc gutters annually up to 300 guilders per kilogram, in order to reduce the use of zinc in the economy.

Recycling scenario
Prices of used materials can change due to changes in market conditions (e.g. costs of disposal of materials) or to government inventions (e.g. subsidies on recycling and deposit-refund systems). In this scenario the government gives a subsidy on the recycling of both used zinc and used pvc. As a result it may become more attractive to collect and recycle the zinc of which zinc gutters are made. The percentage of zinc recycling is arbitrarily assumed to rise linearly with a price increase of used

zinc (p_{zu}), i.e. $80+p_{zu}*10$. Note that the price of used zinc is expressed in guilders per kilogram. The percentage of pvc recycling also rises because of an increase in the price of used pvc as a result of a government subsidy. In 1990 used pvc does not have a positive price because it is cheaper to make new pvc than to recycle used pvc. Later on, when the price of used pvc rises, construction companies will have an incentive to collect the pvc gutters separately for recycling. The percentage of pvc-recycling is assumed to be (70+price of used pvc*10)%. In the model the recycled pvc is assumed to be used for gutters again. In practice, however, the recycled pvc from gutters does not have the same quality as new pvc and therefore cannot be used again for pvc gutters.

Mixed scenario
The mixed scenario combines the two previous scenarios. Prices of zinc gutters, used zinc and used pvc increase, which has an impact on the allocation of the demand and the percentages of recycling for both materials.

In the next section the results for each these scenarios will be presented.

8.6. Results of the Scenario Analysis

Table 8.2 summarizes the simulation results. In the columns of the scenarios the differences for a number of indicator outcomes are shown, with the base scenario outcomes represented with + (-) for a positive (negative) change, ++ for a considerable change, 0 if there is no change and ≈ if there is a very small change.

Table 8.2. Overview of the results of the scenarios.

Scenarios Indicators		Preferences shift	Product charge	Recycling	Mixed
Economic	Demand for gutters	+	+	0	+
	Allocation variable	-	-	0	-
	Net costs	+	+	-	+
Environmental	Recycling zinc (%)	0	0	+	+
	Recycling pvc (%)	0	0	+	+
	Effect on stock zinc ore	++	++	+	++
	Zinc waste in environment	-	-	-	-
	Pvc waste in environment	+	+	-	+
	Galv. steel waste in environment	≈	≈	≈	≈

Results of the base scenario
In the base scenario the exogenously determined variables remain stable over time. During the first 30 (20) years after the base year, the stock of zinc (pvc) gutters is either replaced or disposed of as waste gutters because of the demolition of houses. The number of newly built houses is assumed to remain constant over time. The demand for gutters increases yearly because the number of newly built houses is higher than that of demolished houses and gutters (see line 1 in Figure 8.4). The

allocation of the demand for gutters does not change, the demand for gutters remains at 80% of the total demand (see line 4 in Figure 8.4). The demand for gutters is allocated between zinc and pvc gutters (lines 2 and 3, respectively, in Figure 8.4).

This demand has a substantial influence on the stock of zinc ore as can be seen in Figure 8.5 (line 1). The results of the disposal of zinc, pvc and galvanized steel into the environment are represented in Figure 8.5 by lines 2 to 4. Especially the waste flow of galvanized steel into the environment is significant compared with that of zinc and pvc.

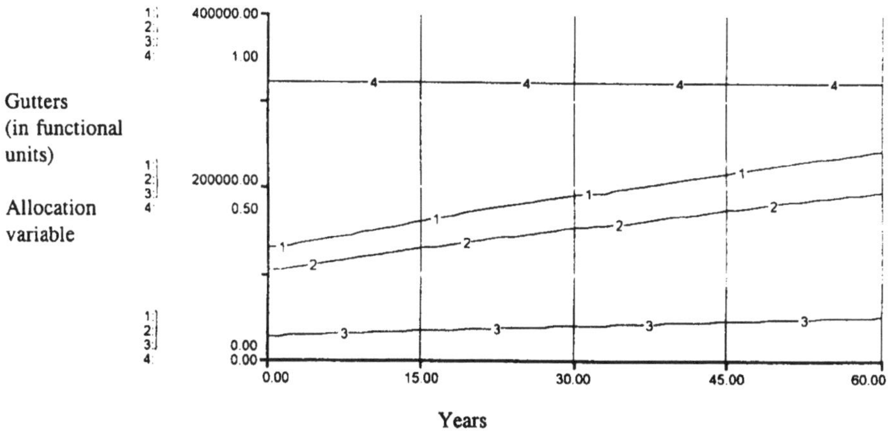

Figure 8.4. The demand for gutters (line 1) allocated to zinc gutters (line 2) and pvc gutters (line 3) by an allocation variable (line 4) in the base scenario.
Note: Lines 1-3: 0-400000, and line 4: 0-1.

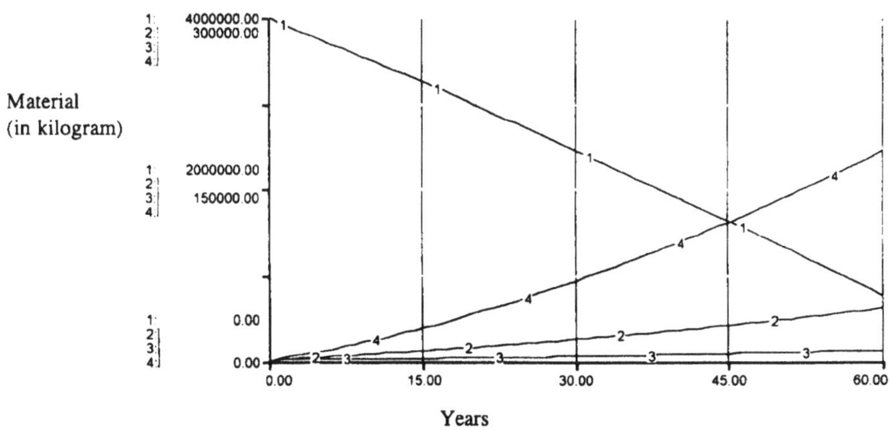

Figure 8.5. The extraction of zinc ore (line 1) and the disposal of zinc (line 2), pvc (line 3) and galvanized steel (line 3) into the environment in the base scenario.
Note: Line 1: [0-4000000], and lines 2-4: [0-3000000]

The prices for the products and for the recycled materials remain stable over time (see lines 1-4 in Figure 8.6). The net costs of satisfying the demand, which are the costs of the demand for gutters minus the revenues from used materials, rise every year because of an increasing demand (see line 5 in Figure 8.6).

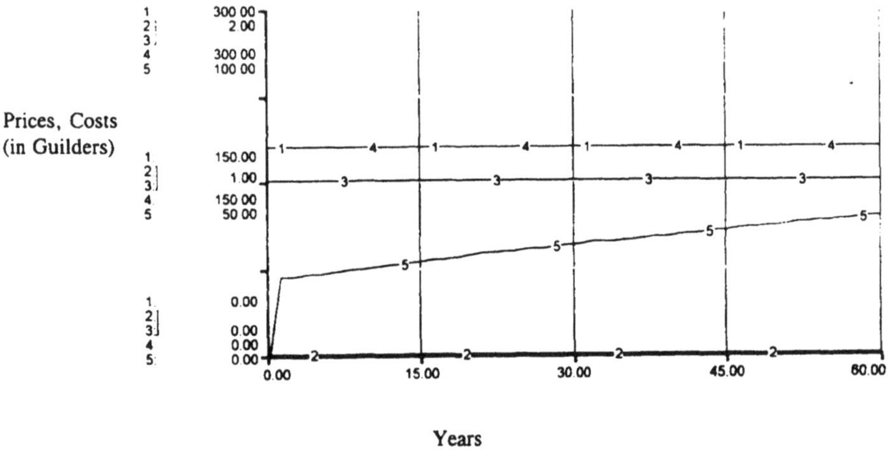

Figure 8.6. Prices of zinc gutters (line 4), pvc gutters (line 1), recycled zinc (line 3) and recycled pvc (line 2), and the net costs of meeting the demand (line 5) in the base scenario.

Note: Lines 1 and 4: [0-300], lines 2-3: [0-2], and line 5: [0-100]

Results of the preferences shift scenario

Because of a growing environmental awareness, preferences are assumed to shift in favour of pvc gutters. Hence, the share of the demand allocated to pvc gutters increases. As a consequence, the number of renovated gutters rises in comparison with the base scenario because pvc gutters have a shorter lifetime.

As an effect of the use of more pvc gutters the net costs over 60 years will rise for two reasons: first, the shorter life-span of pvc gutters means that more pvc gutters are required and therefore this entails higher costs; and second, used pvc does not have any positive impact on total costs as used zinc does.

The total use and disposal of zinc decreases (line 2 in Figure 8.7) while the amount of disposed pvc rises (line 3 in Figure 8.7). The waste of galvanized steel (line 4 in Figure 8.7) decreases because pvc gutters need less galvanized steel for their fastening-pieces, but the use of these pieces is increased because of the shorter lifetime of gutters. In total, the net effect is that the amount of disposed galvanized steel will increase slightly. The amount of zinc ore needed has decreased sharply because of the changed allocation of the gutter demand (line 1 in Figure 8.7).

The market share for gutters decreases from 1990 to 2010, beyond which year the allocation variable remains stable which implies a slight increase in the demand for zinc gutters because of a continuously increasing demand for gutters (see line 4 in Figure 8.8). The demand for gutters (line 1 in Figure 8.8) is allocated between zinc and pvc gutters (lines 2 and 3 in Figure 8.8).

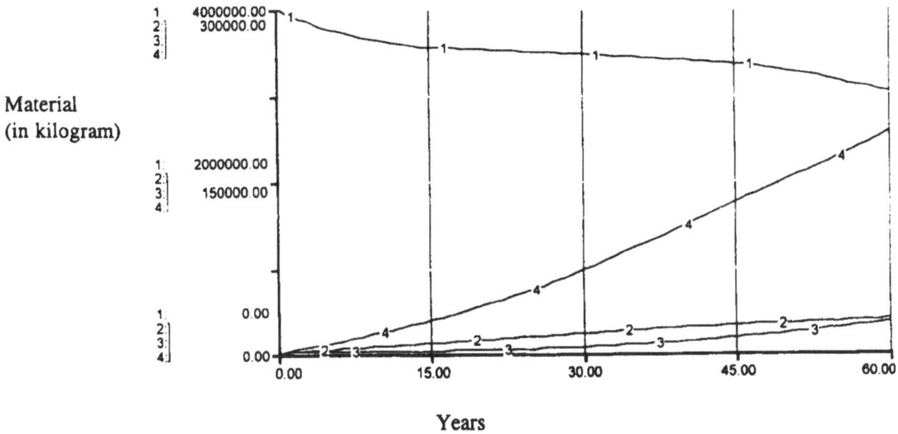

Figure 8.7. The extraction of zinc ore (line 1) and the disposal of zinc (line 2), pvc (line 3) and galvanized steel (line 3) into the environment in the preferences shift scenario.
Note: Line 1: [0-4000000], and lines 2-4: [0-3000000]

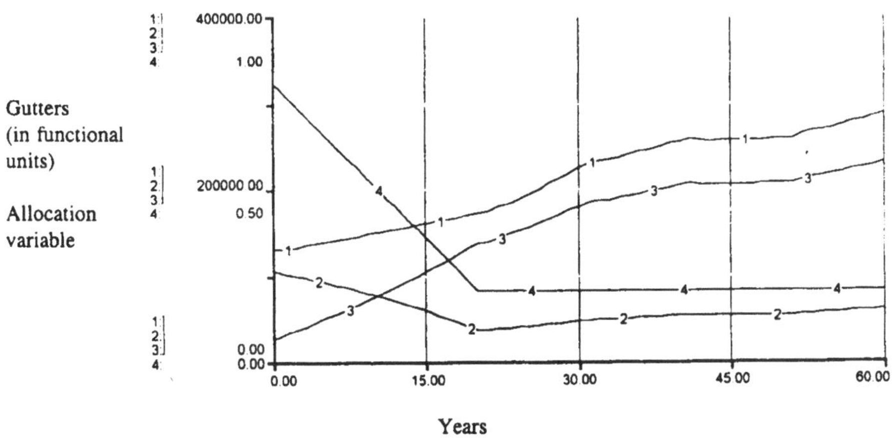

Figure 8.8. The demand for gutters (line 1) allocated to zinc gutters (line 2) and pvc gutters (line 3) by an allocation variable (line 4) in the preferences shift scenario.
Note: Lines 1-3: [0-400000], and line 4: [0-1].

The amount of recycling of zinc is influenced by the decrease of the use of zinc gutters in the first 20 years. The amount of zinc recycling reacts to the demand for zinc gutters with a time-lag of 30 years. In contrast with the demand for zinc gutters, pvc gutters are used more. After 20 years the amount of recycled pvc increases. The costs of meeting the demand rise, because the lifetime of pvc gutters is shorter which means that more such gutters are needed. The revenues from used materials are lower because reused pvc has no positive value. These two effects mean that the net costs rise, but the negative impact on the extraction of materials out of the environment is reduced. The impact of the disposal of materials is twofold: on the one hand, the disposal of zinc and galvanized steel is reduced but on the other hand, the disposal of pvc is increased.

Results of the product charge scenario

In the product charge scenario the price of zinc is increased by a charge imposed by the government. Because of the difference in the prices of pvc and zinc gutters their market shares will change. It is assumed that the percentage of zinc gutters in the total demand decreases from 80% to 20% (see line 4 in Figure 8.9). As a consequence of this, a greater part of the demand for gutters will be met by pvc gutters than by zinc gutters (see lines 2 and 3 in Figure 8.9), which has an impact on the total amount of disposed materials - as in the preferences shift scenario. The impact on the extraction and disposal of materials is roughly the same as in the preferences shift scenario. The demand for gutters shows a more capricious course than in the base and preferences shift scenarios (see line 1 in Figure 8.9 and compare with Figures 8.4 and 8.8).

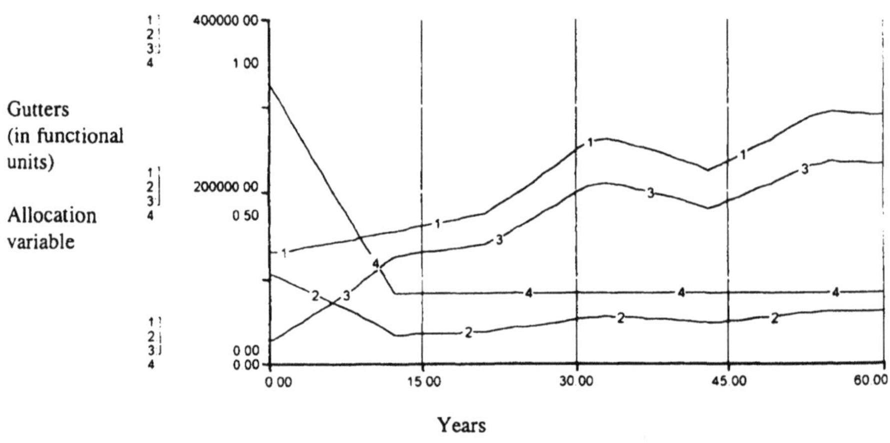

Figure 8.9. The demand for gutters (line 1) allocated to zinc gutters (line 2) and pvc gutters (line 3) by an allocation variable (line 4) in the product charge scenario.
Note: Lines 1-3: [0-400000], and line 4: [0-1].

From years 20 to 30 the demand increases more strongly than before because the pvc gutters that were demanded in the first 10 years are now being renovated. In the first 10 years the share of pvc gutters in the total demand increases strongly. From years 30 to 40 (i.e. 2020-2030) the zinc gutters that were demanded in the first ten years, and the pvc gutters that were demanded 20 years ago, are renovated.

Therefore, there is a sharp increase in the total demand for gutters. However, the cyclical pattern of the demand over time decreases over time, because the impact of the initial change on the market share decreases over time.

The costs of meeting the demand are increased for three reasons: (1) the price of zinc gutters has increased; (2) the pvc gutters have a shorter life-span; and, (3) used pvc does not have a value and therefore does not have an effect on the total costs.

Results of the recycling scenario
In the recycling scenario the prices for used materials are higher which implies that the quantities of recycled zinc and pvc increase (see lines 1 to 4 in Figure 8.10). This scenario does not have an impact on the demand for gutters. Because of the price increases of used zinc and pvc, the recycling percentages are assumed to increase from 80% to 90% for zinc and from 70% to 80% for pvc. Less zinc is extracted which has a small but positive impact on the environment (see line 1 in Figure 8.10 and compare with line 1 in Figure 8.5). The disposal of zinc and pvc has decreased too, compared with the base scenario (lines 2 and 3 in Figure 8.11). The disposal of galvanized steel (line 3 in Figure 8.11) is the same as in the base scenario. The net costs decrease slightly because of the increased prices and amounts of recycled materials.

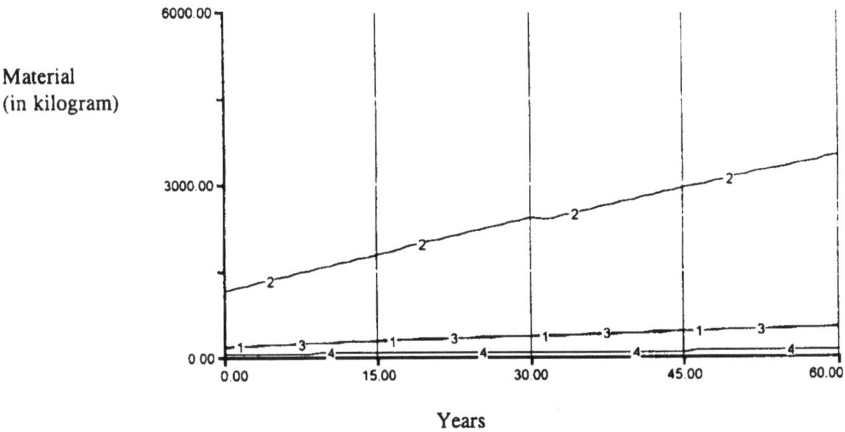

Figure 8.10. The quantities of recycled zinc of demolished houses (line 1) and renovated houses (line 2), and recycled pvc of demolished houses (line 3) and renovated houses (line 4) in the recycling scenario.
Note: Lines 1-4: [0-6000].

Results of the mixed scenario
This scenario combines the assumption of the product charge and the recycling scenarios. In this scenario the prices of zinc gutters, used zinc and used pvc are higher than in the base scenario. The effects on the environment are slightly better than in the product charge scenario because of the small increase of recycled materials. The allocation of the demand for gutters is equal to the product charge

scenario, while the recycling quantities are slightly different compared with the recycling scenario (compare lines 1 to 4 in Figure 8.13 with Figure 8.10). The net costs are slightly lower because of the revenues from selling used materials for recycling purposes. Figure 8.12 shows the net costs (line 5), the prices of zinc gutters (line 4), pvc gutters (line 1), used zinc (line 3) and used pvc (line 2).

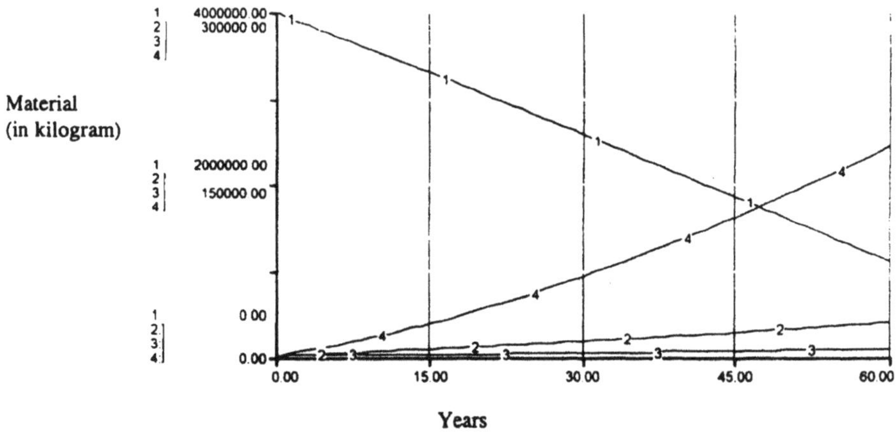

Figure 8.11. The extraction of zinc ore (line 1) and the disposal of zinc (line 2), pvc (line 3) and galvanized steel (line 3) into the environment in the recycling scenario.
Note: Line 1: [0-4000000] and lines 2-4: [0-300000].

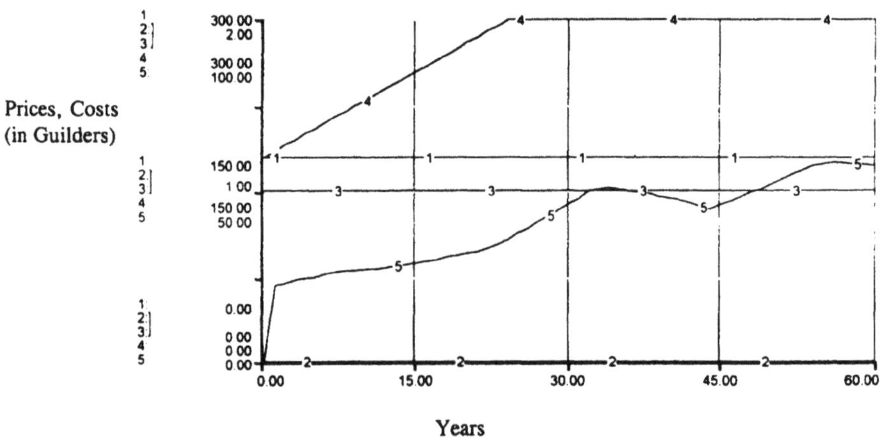

Figure 8.12. Prices of zinc gutters (line 4), pvc gutters (line 1), recycled zinc (line 3) and recycled pvc (line 2), and the net costs of meeting the demand (line 5) in the mixed scenario.
Note: Lines 1 and 4: [0-300], lines 2-3: [0-2], and line 5: [0-100]

Because of the change in the prices for zinc gutters, used zinc and used pvc the quantities of zinc that are recycled decrease steeply after 30 years, while the quantities of recycled pvc increase after 20 years. Afterwards, the quantities of recycled materials show a more stabilized course (lines 1 to 4 in Figure 8.13).

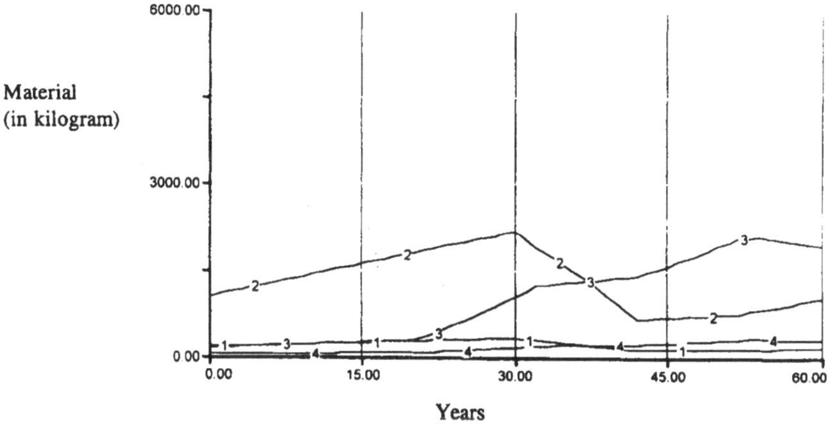

Figure 8.13. The quantities of recycled zinc of demolished houses (line 1) and renovated houses (line 2), and recycled pvc of demolished houses (line 3) and renovated houses (line 4) in the mixed scenario.
Note: Lines 1-4: [0-6000].

8.7. Conclusions

The demand for a service can be met by various products made of different materials. Material flow models capture these. Most models of material flows are based on the material balance principle. However, they ignore the role of products and economic forces operating on materials and products. The concept of an M-P chain takes into account the substitutability and complementarity of products and materials. Therefore, links between materials and products are considered.

In this study a simple dynamic simulation model for M-P chains for gutters was presented that was intended: (1) to reflect the importance of products and processes for the analysis of flows of materials; (2) to track the impact of economic and government policy variables on M-P chains; and, (3) to include dynamic aspects such as accumulation and delays.

The flows of materials and products through the economy may alter as a result of changes in prices. It is important to look simultaneously at the material and product flows because they are strongly linked. The use of M-P chains as units of analysis is justified since particular physical as well as economic aspects are incorporated.

A model of a particular M-P chain has been developed based on some characteristics of a system with two alternative products to satisfy a specific need: zinc and pvc gutters to meet the demand for gutters. The environmental impacts of this M-P chain are related to zinc ore depletion and waste flows of zinc, galvanized steel and pvc into the environment. The economic aspects were limited here to the net costs of satisfying the demand for this service. The model includes dynamic processes: for example, the preferences are assumed to shift gradually over time, and the product charge rises over time. Accumulation and delayed effects are explicitly taken into account: for example, a shift from zinc to pvc rain gutters in the preference shift scenario affects the quantities of zinc and pvc that are disposed of with a delay of many years. These dynamic processes cannot be analysed with a static analysis.

The results of the simulations show that the higher prices of zinc gutters and the preferences of consumers for a certain type of gutter have a strong impact on the use of materials. An increase in the prices of used materials can raise the percentages of these materials which are recycled. However, the influence of a price increase of used zinc is not great because the percentage recycled is high already. But a change in preferences for this type of gutter has a significant influence on the use of zinc and pvc. To summarize, apart from the physical flows of materials and products, it is important to look at the impact of economic and government variables on flows.

There are several possibilities to extend this model with other economic characteristics. Some further research topics are: elaboration of the dynamics of the M-P chains, e.g. regarding technological developments in the recycling of galvanized steel; and the extension of the M-P chains with more products or materials, e.g. by taking other zinc products into account. Data availability and the uncertainty about future events may prove to be serious obstacles in using M-P chain models to explore future developments in relation to material flows. Nevertheless, they do present an important approach to study some of the main forces driving material flows.

Appendix 8. The Equations of the Model with Explanation

Submodel for rain gutters in housing

1) init g_acc = 2,660,560
Initially the number of gutters in the economy is set at 2,660,560. This is less than that estimated by Fraanje and Verkuijlen (1996) but it is more consistent with the estimates of the renovation and demolition of gutters. The stock of gutters equals the stock of last year plus the new gutters minus the gutters in demolished and renovated houses.
2) g_dem_in = pg_dem_in+zg_dem_in
The total number of demolished gutters is the sum of the demolished pvc and zinc gutters.
3) g_ren_in = pg_ren_in+zg_ren_in
The total number of renovated gutters is the sum of the renovated pvc and zinc gutters.
4) g_new_in = 75,000+g_ren_in
Every year 75,000 new gutters are needed for the construction of new houses in addition to the number of renovated gutters.

Submodel for zinc rain gutters

5) init zgacc = 2,660,560*0.8
The initial stock of zinc gutters is 80% of all gutters (see equation 1).
6) zgnew_in = g_new_in*r
The amount of new zinc gutters are a part, r, of the new gutters.
7) i) r=0.8
In the base scenario the percentage of the gutter demand which is satisfied by zinc gutters is 80%.
ii) r=max(0.8-0.03*t,0.2)
In the preference shift scenario the percentage of the gutter demand which is satisfied by zinc gutters is r=max (80-3*t,20)%, which means that the part of the demand which is allocated to zinc gutters decreases over time until a certain minimum.
iii) r = max(0.8-0.01*(price_zn_gut-price_pvc_gut), 0.2)
The percentage of the demand which is satisfied by zinc gutters depends on the prices of both zinc and pvc gutters.
8) zg_dem_in = if t<31 then (48,832+1,812*t)*0.12 else dummy1*0.12
zg_ren_in = if t<31 then (48,832+1,812*t)*0.88 else dummy1*0.88
dummy1 = delay(zgacc_in,30)
In the first 30 years the initial number of zinc gutters is either renovated or demolished. The amount of gutters that are renovated in 1990 is derived from Fraanje and Verkuijlen (1996). In the following years more zinc gutters are renovated or demolished because the construction of houses has risen in the period from the 1960s to the 1990s. After 30 years the number of demolished and renovated gutters is set equal to the new gutters with a time-lag of 30 years, i.e. the new gutters of 1990 are demolished or renovated in 2020. The 30-year time-lag is taken as an average lifetime of all gutters (Fraanje and Verkuijlen, 1996).

Submodel for pvc rain gutters

9) init pvc_g_acc = 2,660,560*0.2
Initially 20% of the gutters are made from pvc (see equation 1).
10) init pg_new = 75,000*0.2
Initially 20% of the new gutters are allocated to pvc.
11) pg_new_in = g_new_in*(1-r)
A part of the demand for gutters is allocated to pvc.
12) pg_dem_in = if t<21 then (12,208+679.6*t)*0.12 else dummy2*0.12
pg_ren_in = if t<21 then (12,208+679.6*t)*0.88 else dummy2*0.88
dummy2 = delay(pg_acc_in,20)
In the first 20 years the initial amount of pvc gutters is either renovated or demolished. The

amount of gutters that are renovated in 1990 is derived from Fraanje and Verkuijlen (1996). In the following years more zinc gutters are renovated or demolished because the construction of houses has risen in the period from the 1960s to the 1990s. Afterwards, the number of renovated and demolished houses are set equal to the new pvc gutters of 20 years ago. This 20-year time-lag is equal to the average lifetime of a pvc gutter.

Submodel for fastening-pieces

13)init accfastp = 2,660,560*0.2*0.0168

Initially 20% of the gutters are made from pvc. For every pvc gutter 16.8 kilograms (0.0168 tonnes) of galvanized steel is needed to make the fastening-pieces.

14) init accfastz = 2,660,560*0.8*0.0258

Initially 80% of the gutters are made from zinc. For every zinc gutter 25.8 kilograms (0.0258 tonnes) of galvanized steel is needed to make the fastening-pieces.

15) accfp_in = g_acc_in*(1-r)*0.0168

The number of fastening-pieces of galvanized steel for pvc gutters in the economy depends on the demand for pvc gutters. Note that r is the part of gutters which are made from zinc.

16) gal_w1 = (pg_ren_w_in+pg_sloop_in)*0.0168

The quantity of galvanized steel which goes to the environment as waste depends on the number of pvc gutters which are disposed of.

17) accfz_in = g_acc_in*r*0.0258

The number of fastening-pieces of galvanized steel for zinc gutters in the economy depends on the demand for zinc gutters. Note that r is the part of gutters which are made from zinc.

18) gal_w2 = (zgren_w_in+zgsloop_in)*0.0258

The quantity of galvanized steel which goes to the environment as waste depends on the number of zinc gutters which are disposed of.

Submodel for zinc

19) init new = (75,000+72,000*0.75)*0.8*0.0296+100

The zinc which is needed for the zinc gutters is produced. Production waste of 100 tonnes is included. For one zinc gutter 0.0296 tonnes of zinc ore is needed.

20) init zacc = 2,660,560*0.8*0.0296

Initially there are 2,660,560*0.8*0.0296 tonnes of zinc.

21) zinc_out = zgnew_in*0.0296

The amount of zinc needed for the gutters equals the number of gutters multiplied by the amount of materials in tonnes per gutter.

22) z_dem_in = zg_dem_in*(0.0296-0.00011*30)

The amount of demolished zinc equals the number of demolished zinc gutters multiplied by the amount of materials in tonnes per zinc gutter.

23) z_ren_in = zg_ren_in*(0.0296-0.00011*30)

The amount of renovated zinc equals the number of renovated zinc gutters multiplied by the amount of materials in tonnes per zinc gutter.

24) i) z_rec1_in = 0.9*z_dem_in

The amount of recycled zinc from demolition depends on the price of used zinc and the amount of demolition waste. In the base scenario the price of recycled zinc is 1 which means that 90% of the demolished zinc-waste is recycled.

ii) z_rec1_in = 0.8+price_reczn-0.1

The percentage of zinc-recycling is assumed to be (80+price of used zinc*10)% for both the renovation and the demolition waste in the recycling-policy scenario.

25) i) z_rec2_in = 0.9*z_ren_in

The amount of recycled zinc from renovation depends on the price of used zinc and the amount of renovation waste. In the base scenario the price of used zinc is 1 which means that 90% of the renovated zinc-waste is recycled.

ii) z_rec2_in = 0.8+price_reczn-0.1

The percentage of zinc-recycling is assumed to be (80+price of used zinc*10)% for both the demolition and the renovation waste in the recycling-policy scenario.

26) rec_out = reczn

The recycled zinc goes directly into the zinc-stock.

27) zleach_in = 0.00011*zgacc

Every year a part of the zinc gutters is leached out.

28) z_envl_in = zleach_in

All the leached zinc goes directly to the environment.

29) z_env2_in = z_dem_in-z_recl__in

The part of the demolition waste which is not recycled is disposed of into the environment.

30) z_env3_in = z_ren_in-z_rec2_in

The part of the renovation waste which is not recycled is disposed of into the environment.

31) z_env4_in = 100

The production waste is 100% every year.

32) r (see equation 7)

Submodel for pvc

33) init p_acc = 0.0161*0.2*2,660,560

Initial amount of pvc in the economy.

34) p_rec_out = p_rec

Recycled pvc does not have the same quality as new zinc. It cannot be used for pvc gutters again and therefore it goes to another sector.

35) i) p_recl_in = 0.7*p_ren_in

The quantity of recycled pvc from renovation depends on the price of used pvc and on the amount of renovated pvc-waste. In the base scenario the price of used pvc is 0.

ii) p_recl_in = 0.7+price_recpvc*0.1

The percentage of pvc-recycling is assumed to be (70+price of used pvc*10)% for both the demolition and the renovation waste in the recycling-policy-scenario.

36) i) p_rec2_in = 0.7*p_dem_in

The quantity of recycled pvc from the demolition of houses depends on the price of used pvc and on the amount of demolished pvc-waste. In the base scenario the price of used pvc is 0.

ii) p_rec2_in = 0.7+price_recpvc*0.1

The percentage of pvc-recycling is assumed to be (70+price of used pvc*10)% for both the renovation and the demolition waste in the recycling-policy scenario.

37) p_dem_in = pg_dem_in*0.0161

p_ren_in = pg_ren_in*0.0161

p_new_in = pg_new_in*0.0161

The amount of pvc gutters in the different stages of the chain can be converted into tonnes by multiplying by 0.0161.

38) p_envl_in = p_dem_in-p_rec2_in

The part of the demolished pvc-waste which is not recycled is disposed of into the environment.

39) p_env2_in = p_ren_in-p_recl_in

The part of the renovated pvc-waste which is not recycled is disposed of into the environment.

Submodel for ores

40) init fe_ore = 4,000,000

Initially the iron-ore stock is arbitrarily set at 4 million tonnes.

41) init zinc_ore = 4,000,000

Initially the zinc ore stock is arbitrarily set at 4 million tonnes. Every year zinc is extracted for zinc gutters and for the production of galvanized steel to make two types of fastening-pieces.

42) fepvc_in = accfp_in*1.44

One tonne of galvanized steel for a pvc-fastening-piece needs 1.44 tonnes of iron-ore.

43) fezn_in = accfz_in*1.44

One tonne of galvanized steel for a zinc-fastening-piece needs 1.44 tonnes of iron-ore.

44) znpvc_in = accfp_in*1.67

One tonne of galvanized steel for a pvc-fastening-piece needs 1.67 tonnes of zinc ore.

45) znzn_in = accfz_in*1.67

One ton of galvanized steel for a zinc-fastening-piece needs 1.67 tonnes of zinc ore.

46) zinc_ore_out = (zinc_out-z_rec_out)*22.3

The zinc ore which is used for zinc gutters is the amount of new produced zinc multiplied by the amount of zinc ore needed for one tonne of zinc, i.e. 22.3 tonnes.

Submodel for costs and prices

47) net_costs_in = (zgacc_in*price_zn_gut+pg_acc_in*price_pvc_gut-1000*(reczn*price_reczn-p_rec_out*price_recpvc))/1,000,000

The net costs of satisfying the demand for gutters is equal to the price of zinc gutters multiplied by their demand plus the price of pvc gutters multiplied by their demand minus the revenues from used zinc and used pvc. These revenues are equal to the price of used zinc multiplied by the amount of recycled zinc and the prices of used pvc multiplied by the amount of recycled pvc. The revenues have to be multiplied by 1000 because the price is given per kilogram and the amount in tonnes. The net costs are in millions of Dfl per year which means that the right-hand side has to be divided by 1,000,000.

48) price_pvc_gut = 180

In the base case the price of a pvc gutter is Dfl 180 and does not change over time (Tauw Milieu, 1994).

49) i) price_recpvc = 0

The price of used pvc is Dfl 0 per kilogram.

ii) price_recpvc = 1

In the recycling and mixed scenario the price of used pvc is Dfl 1 per kilogram.

50) i) price_reczn = 1

The price of used zinc is Dfl 1 per kilogram (*Recycling*, 1994).

ii) price_reczn = 2

In the recycling and mixed scenario the price of used zinc is Dfl 2 per kilogram.

51) i) price_zn_gut = 180

In the base case the price of a zinc gutter is Dfl 180 and is assumed to remain stable over time (Tauw Milieu, 1994).

ii) price_zn_gut = min(180+charge,300)

In the product charge scenario the price for zinc gutters is initially Dfl 180, but the government imposes a charge which increases the price of a zinc gutter.

charge= min(5*t,120)

The charge will be raised by Dfl 5 every year, up to a maximum of Dfl 120.

CHAPTER 9

A DYNAMIC ANALYSIS AND EVALUATION OF WINDOW FRAMES[1]

9.1. Introduction

The present chapter is concerned with the study of the M-P chain of window frames used in houses. This has been chosen as a case study because window frames are used on a large scale, are durable goods, and the respective chain creates a variety of different environmental pressure impacts, notably via the use of various types of materials and energy which causes impacts on resources use and generation of waste (Holmes, 1990; VROM/DGM, 1993; VROM, 1995). A number of different chain management instruments and environmental and economic indicators will be linked in a dynamic model. These indicators are, among other things, raw materials depletion, water pollution, costs and prices. This dynamic modelling approach makes it possible to study the impact of different policy instruments on the choice of specific types of window frames and subsequently on material flows and related environmental impacts over time. Numerical simulation is employed to examine time patterns for the relevant indicators under different policy and external scenarios.

The dynamic M-P chain approach adopted here adds to these studies a more integrated analysis of decision and physical processes, based on a dynamic representation of product and material flows, accumulation of materials in products through a vintage approach, and linkages of various environmental pressure indicators to material flows.

In Section 9.2 the concept of M-P chains is discussed in more detail. Section 9.3 reviews studies presenting information on the environmental and economic impacts of the use of window frames in houses. In Section 9.4 the dynamic model for the M-P chain of window frames is presented. Section 9.5 contains a description of policy instruments and scenarios considered in the policy analysis. The results of the scenario analysis with the window frame chain model are presented in Section 9.6. Concluding remarks are provided in a final section.

9.2. Analysis of M-P Chains for Several Products

The demand for a service can generally be met by different products. When this is the case, the respective products are usually imperfect substitutes. This implies that as the price of one product increases, the demand for another increases. As an

[1] This chapter is a slightly revised version of an article published in *Ecological Economics* (Kandelaars and Van den Bergh, 1997a).

example, a bicycle and a car can both fulfil the service of moving people or freight from one place to another, but do not perfectly substitute for one another. Differences between (imperfect) substitute products that are of particular importance in the context of environmental policy relate to the materials that these products are composed of, and the environmental impacts associated with, their production and use. Here it is argued that an M-P chain framework allows one to consider all these elements in a systematic and consistent manner.

Figure 9.1 shows the general structure of an M-P chain. The term 'chain' refers to the sequence of activities associated with a particular service, which includes processes falling into general categories like extraction, production (including physical and chemical transformation and assemblage), consumption, materials and product recycling, and treatment, incineration and dumping of waste. An M-P chain thus combines a representation of material flows from environmental sources to environmental sinks with that of materials cycles and accumulation of materials in (durable) products (or capital) in the economy. At one end of the flow there are the waste materials derived from discarded products which can be dumped or recycled. The use of waste materials can be influenced by changing the relative prices of dumping or recycling, for instance, via a levy on dumping or by subsidizing recycling. At the other end of the chain is the extraction of virgin materials, serving as inputs to production and sometimes consumption processes. The choice between new (raw) or recycled materials for the production of goods depends on relative prices, the relative quality of virgin versus recycled materials, and on available or implemented technologies (related to product design). It should be noted that when considering a specific chain, the output and input of recycled materials are not necessarily equal, because recycled materials can go to the production of other types of products outside the chain, and the input of recycled materials can originate from the waste of products outside the respective chain. The quality and the prices of recycled materials are not necessarily equal, and indeed will often differ, and price differences are expected to be related to such quality differences.

An important category of goods are durables, such as (parts of) houses, cars, and various devices used in household activities (for cooking, cleaning, cooling, etc.). These usually use large amounts of particular materials (such as metal alloys). The durability of these goods implies that waste flows from their consumption will generally arise a long time after decisions related to their production and consumption are made. This means that planning of both recycling and waste treatment may be difficult. For instance, in the case of the window frames considered in the application later on, it takes thirty years until the materials they incorporate become potential waste. By that time, relevant technological profiles of production or waste treatment may well have completely changed.

Formal models of an M-P chain can provide information on a number of relevant issues in terms of long-term environmental policy goals. This may include information about: the reduction in materials use, in residuals emission and in waste generation; the shift from non-renewable to renewable and recyclable materials; and the reduction in energy use. Some of these issues can be studied by way of a static model. This may be either a static equilibrium model with policy instruments, or a static optimization model (see Chapter 6) where costs of given supply are minimized. These models do not take into account various temporal issues relevant over both short and long run periods: dynamic processes like adoption of new technology, product evolution (including changes in materials composition) and shifts in demand and preferences. Shorter run processes relevant in an M-P chain context are the accumulation of substances in (durable) consumption and capital goods, formation of vintage structures of durable goods in-use, where each vintage is associated with a unique materials composition of the respective good; and delays between subsequent activities in the M-P chain. These various phenomena clearly indicate that a dynamic model is essential to address questions of long-term development of M-P chains, and related to it, chain management oriented toward environmental sustainability. The following section illustrates some of these dynamic phenomena in a dynamic modelling exercise for window frames.

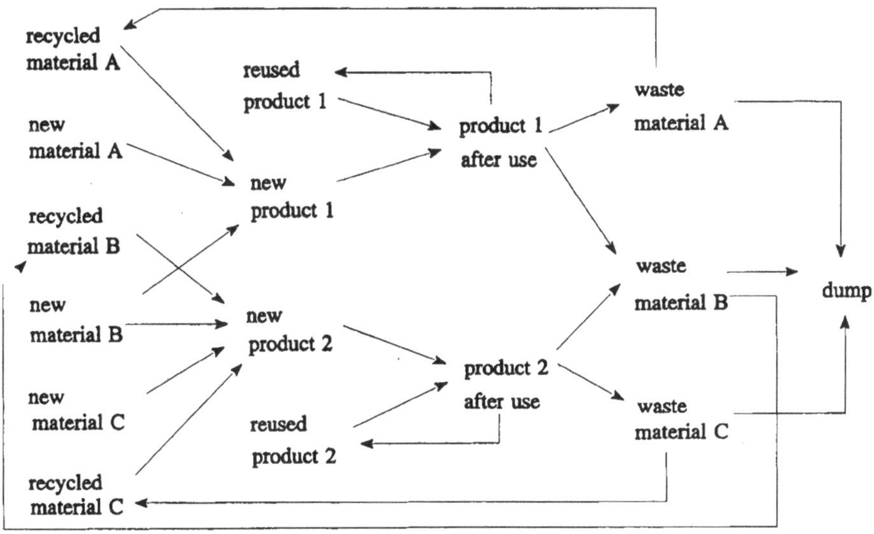

Figure 9.1. A material-product chain for two products

9.3. Overview of Studies on Window Frames

In the remainder of this paper we will be concerned with a dynamic M-P chain analysis of the window frame service. A range of alternative products is available for this service, based on alternative materials. The intended M-P chain analysis requires data regarding the material, environmental and economic characteristics of these products. This section summarizes several studies that have provided these data. The environmental and economic impacts of the different window frame types have been made comparable by defining a functional unit as 1 square meter window frame (adopted from Lindeijer *et al.*, 1990). Table 9.1 lists the relevant studies of window frames and displays the material, environmental and economic characteristics that have been taken into account by each. The materials include hardwood, pinewood, pvc and aluminum. The environmental effects embody, among other things, raw materials use, energy use and emissions. The economic characteristics cover purchase, maintenance, lifetime, recycling and dumping costs.

Holmes (1990) investigates why window frame producers in the United Kingdom have chosen certain materials for their products. A survey showed that the most important factors influencing the use of materials in window frames are consumer preferences, specifically related to the type of house (new or old). Furthermore, it was assessed that changes in the prices of materials have had a significant impact on the price of the window frames, but the latter was found not to influence the use of the window frames very much. The reasons for this may be that consumers do not only take into account the price of a window frame in their decisions, but also the additional costs associated with installation and future maintenance. These latter costs are quite different among the alternative types of window frame (associated with different materials use).

Table 9.1. Materials, environmental and economic characteristics of studies of window frames.

Characteristics Study	Materials	Environmental	Economic
Holmes (1990)	Aluminium, pvc, wood, steel		Choices for material by producer
Hendrix and Martens (1990)	Aluminium, pvc, hardwood, pinewood (2 types)	Various environmental indicators	Costs of the entire life cycle of product
Lindeijer *et al.* (1990)	Aluminium, pvc, hardwood, pinewood, steel	Various environmental indicators	
VCNI (1991)	Aluminium, pvc, wood	Various environmental indicators	Recycling and dumping costs, substitution and recycling
Hoefnagels *et al.* (1992)	Aluminium, pvc, hardwood, pinewood, steel	Various environmental indicators	Recycling, reuse and design
Fraanje *et al.* (1992)	Larch wood	Environmental problems of using larch wood	

Hendrix and Martens (1990) investigate which types of window frame are preferable from an economic and an environmental policy perspective. They conclude that pinewood window frames painted with natural paint are environmentally preferable, that is pinewood has the best score over various environmental aspects, in comparison with window frames made of hardwood, aluminum, pvc, or pinewood coated with acrylic paint. The cumulative costs over the life time of 50 years, including maintenance costs, have been estimated with a discount factor of 3%. These show that window frames made of pinewood and painted with acrylic paint are the cheapest. In this case it appears that there may be a conflict between choosing the product type based on a purely environmental or a purely economic criterion.

Lindeijer *et al.* (1990) analyse the environmental effects of five types of window frames based on a life cycle approach, taking into account production, use and waste. Either the product can be reused or parts of the product can be recycled. The window frames are compared in terms of energy use, use of hardwood, pollution of water, pollution of air, acidification, and waste. The results show that a single, environmentally most-preferable window frame cannot be determined.

Hoefnagels *et al.* (1992) is a continuation of Lindeijer *et al.* (1990). It examines the environmental impacts of different designs, and of recycling and reuse alternatives for five types of window frames. The results show that several improvements are possible to reduce the environmental impacts of window frames, such as recycling and redesign.

VCNI (1991) examines the environmental and economic effects of reducing the environmental impacts of pvc window frames by means of (i) recycling pvc, (ii) substituting pvc window frames by aluminum ones, or (iii) substituting pvc by pinewood window frames. The economic impact of the three possible improvements is negative and the environmental impact is only positive for option (i).

In Fraanje *et al.* (1992) the bottlenecks, and possible elimination of these, related to the use of larch wood (a type of pinewood) for window frames are examined. It is concluded that in addition to recycling, reuse and redesign, use of new and other materials is a realistic option to reduce environmental impacts.

For the M-P chain analysis as presented in the following sections information about environmental impacts was obtained from Lindeijer *et al.* (1990) and Hoefnagels *et al.* (1992), and about economic characteristics and parameters from Hendrix and Martens (1990). The data from the above studies provide an empirical database for setting parameter values in the dynamic model developed in the next section, as well as for formulating the scenarios in Section 9.5.

9.4. A Dynamic Model with Economic and Environmental Indicators

A dynamic model is presented which can be used to analyse the links between materials and products flows for the M-P chain of window frames. The model consists of three interdependent levels: a physical-material level (measurement in kilograms); an economic level (measurement in monetary units, namely Dutch guilders); and an environmental pressure level (containing mixed units). The material level consists of technological data on window frames, e.g. the amount of kilograms of material per functional unit for each type of window frame. Prices and preferences determine the use of products and materials which in turn determines environmental pressure impacts. Policies can be imposed to reduce the environmental damage by changing prices, preferences, recycling percentages or technologies. Prices can be influenced by charges on a specific type of window frame. Thus the consumer price equals the producer price plus the charge. Preferences can be changed by providing information about environmental pressure associated with using particular types of window frames. Appendix 9 contains details about the initial conditions and specific equations of the model.

For each year, the demand for window frames is determined by the number of new and renovated houses, partly based on forecasts by the Dutch Ministry of Housing (VROM, 1992). In addition, the demand in the model contains an endogenous part, based on the average price of the four types of window frames. The demand for window frames for renovated houses depends on the number of renovated houses, the average price of window frames and time. If the average cost is higher than 4600 guilders, the demand for renovated houses decreases every year by 5% until the limit of 75%. As no long-term forecasts are available for most units, specific assumptions are made about the distribution mechanism overall demand, costs and prices. The equations are not based on an econometric analysis, i.e. there are no statistical estimates of future relationships depending on past, observed relationships. The main reason is that too little variation in relative prices has occurred, as a result of which accurate estimates are impossible. The distribution of the total demand between the four types of window frames is assumed to depend on the prices of each type of window frame, and on consumer preferences. These preferences are furthermore regarded to differ between owners of new and of renovated houses (cf. Fraanje *et al.*, 1992). Table 9.2 presents the distribution of the total demand in 1990 for new and renovated window frames over each type of material in percentages.

Table 9.2. The distribution of the demand for window frames over the materials in 1990, the Netherlands.

Materials	Window frames for new houses	Window frames for renovated houses
Hardwood	62%	39%
Pinewood	8%	7%
Pvc	25%	36%
Aluminium	5%	18%

Figure 9.2 summarizes the structure of demand formation in the model. The demand D_i for product type i (i=1,...,4) can be formulated to depend on total demand for window frames (D) via weights a_i (summing to 1) as follows (all variables have a suppressed time index).

$$D_i = a_i D \quad (1)$$

The preferences of the consumers for a window frame of type i are represented by an index w_i. This index is variable and takes a higher value when the window frame of type i become less attractive (in some scenarios). zet p_i denote the price of a window frame of type i. To calculate the part of the total demand that is allocated to a window frame of type i, the weights a_i are defined as follows.

$$a_i = \frac{\sum_k p_k w_k}{p_i w_i} \quad (2)$$

so that

$$\frac{\partial a_i}{\partial p_i} < 0 \ , \ \frac{\partial a_i}{\partial w_i} < 0 \ , \ \frac{\partial a_i}{\partial p_k} > 0 \ , \ \frac{\partial a_i}{\partial w_k} > 0 \text{ for } i \neq k \tag{3}$$

The two latter partial derivatives show that (2) formalizes that the products are i imperfect substitutes. When the price of one product increases the relative demand for the other products increases. Furthermore, when one product becomes less attractive (i.e. w_k increases) the relative demand for the other products rises.

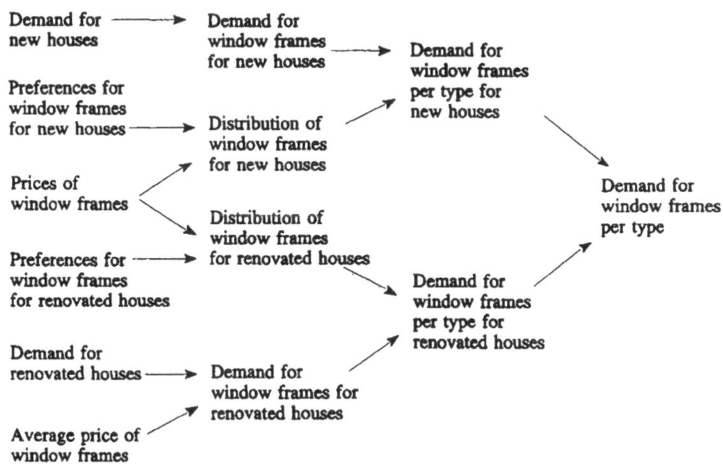

Figure 9.2. The determination of the demand for window frames per type.

For the scenario analysis with the model the lifetime of a window frames is set at 30 years, the base year is 1990 (so that the model uses information from 1960 onwards), and the time horizon is 60 years. The latter allows for dealing with two periods of 30 years in which the complete replacement of the stock of vintages of window frames takes place, and the effects of changes in the vintage structure on environmental pressure indicators can be examined to their full extent. This means that a relevant period for developing scenarios is 1960-2050.

There are three main groups of economic agents involved in the M-P chain of window frames: consumers, producers and a regulator (e.g. a local or national government). The regulator imposes levies on materials and charges on products, determined within each scenario (see next section). The model is partial because it does not include the return of revenues to producers or consumers. A general equilibrium type of approach would be required to deal with this, but the combination of dynamics and a completely closed model would make the analysis far too complicated (see Chapters 7 and 10 on general equilibrium). As one may also expect that the indirect effects of recycling of revenues on the (economic and environmental) performance of the (partial) M-P chain of window frames can be neglected, such a complicated approach is not adopted here.

The window frames are produced with the input of new and recycled materials. The producer's choice regarding these depends on their relative prices and techniques available. These can be altered by imposing levies on new materials or subsidizing the use of recycled materials. Shifts in process techniques used can be stimulated by investment subsidies. The producer pays the price plus levies on the input of new and recycled materials, and the levy on waste materials goes to recycling and dumping. It is assumed that producers are able to supply exactly the number of window frames required to meet the demand.

The choice of instruments and the phase in the M-P chain where they are implemented determine the economic and environmental impacts of these instruments on the chain. For instance, a levy on the use of new materials not only affects the amount of new materials used, but also the amount of recycled materials, and eventually the choices made by final consumers.

The consumer's average cost per window frame is based on the sum of producer prices and charges on window frames. These average costs serve as an indicator of relative economic performance. Serving as a second economic indicator is the total value of charges on products paid by the consumers. After consumption the materials can be either dumped (e.g. in landfills) or recycled to production. That choice depends on the relative costs of dumping and recycling of materials. The price of dumping is exogenous and set locally or nationally.

Additional performance indicators include material and environmental pressure measures. These include raw materials depletion (in depletion units), energy use (in Mjoules), acidification (in acidification units), global warming (in global warming potential, as defined by the IPCC scientific assessment in 1990), water pollution (in cubic metres of potentially polluted water), and solid waste (generated in kilograms). These performance indicators are taken from Hoefnagels *et al.* (1992), and are chosen because they are both relevant and measurable. Noise, land use and other effects, for instance, are not taken into account because they are difficult to assign to specific products or materials.

9.5. Scenarios and Policy Instruments

Based on the dynamic M-P chain model discussed in the previous section, the relationship between the use of materials and the use of products is studied under different scenarios. Three scenarios are considered: (1) a reference scenario; (2) a policy focusing on depletion of raw materials; and, (3) a policy focusing on water pollution. The second scenario is selected because the use of hardwood in houses, including window frames, is regarded as an important cause of tropical deforestation, and has given rise to many policy discussions about discouraging its use. The policy aimed at the reduction of water pollution is interesting as it is mainly related to aluminum and pvc use, for which recycling seems to be a realistic solution.

In the reference scenario no specific policy instruments are implemented for window frames and the materials of which they are made. Of course, any hidden policies presently implemented, which influence the prices of materials and products, are implicit in this scenario as the data did not make it possible to correct for them. The distribution of the total demand over the four types of window frames remains the same over the whole simulation period as in the base year 1990 (Fraanje *et al.*, 1992). The demand for window frames for new and renovated houses is distributed over hardwood, pinewood, pvc and aluminum. Under this scenario no recycling occurs so that all waste is either dumped or incinerated.

Under the second scenario the policy maker wishes to reduce the use of hardwood. Two types of instruments can be used for this purpose: namely, imposing a levy on the dumping of hardwood, and imposing a charge on the use of window frames made of hardwood. Another option is putting a levy on the use of new materials in order to stimulate recycled hardwood for the production. However, this option is not realistic, because the quality of recycled hardwood is much lower than that of new hardwood.

In the third scenario the reduction of water pollution caused by the use of aluminum and pvc in window frames is considered. This is related especially to the dumping of materials after use. Three instruments can be used for this purpose: a levy on the dumping of both materials to reduce the dumping of aluminum and pvc after use; a levy on the use of new aluminum and pvc to encourage producers to use more recycled materials; and a charge on the use of aluminum window frames.

The levies on materials and the product charges are imposed from 1995 onwards. It is assumed that there are no restrictions on the use of recycled materials up to a certain technological limit. For the use of recycled aluminum and for the recycling of aluminum this limit equals 90%. For hardwood recycling this static limit is 75%. Note that recycled hardwood cannot be used for the production of window frames for quality reasons. The technological limit for the use of recycled pvc and of pvc recycling is assumed to change over time from 20% in 1990 to 80% in 2010 (based on VCNI, 1991).

Table 9.3. Policy instruments analysed in the different scenarios.

Policy instruments	Material on which policy is imposed	
	Scenario 2	Scenario 3
A levy on the dumping of materials	Hardwood	Aluminium and pvc
A levy on the use of new materials	-	Aluminium and pvc
A levy on the use of specific products	Hardwood	Aluminium
Information on specific products	Pinewood is more attractive	Pinewood is more attractive

Note: Explanations of scenarios in text.

In addition to the levies and charges, in both policy scenarios the instruments of education and information are included to allow for a direct shift of consumer preferences in favour of pinewood window frames.[2] Table 9.3 summarizes both policy instruments used in each of the two scenarios.

9.6. Results of the Scenario Analysis

The programming of, and dynamic simulation with, the model presented in Sections 5 and 6 has been performed in STELLA II.[3] In the reference scenario (Scenario 1) the preferences of the consumers do not change over time. No recycled materials are used for the production of window frames, and after discarding the products all materials are disposed of. The demand for new and renovated houses in the reference scenario is presented in Figure 9.3. The demand for new houses decreases until 2004, after which the demand remains stable at a higher level. The demand for renovated houses decreases over time.

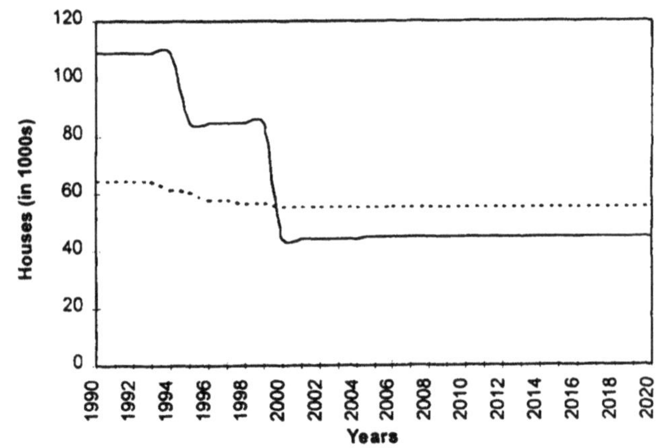

Figure 9.3. The demand for new (solid line) and renovated houses (broken line).

Under the second scenario a levy is put on the dumping of hardwood and on the use of hardwood window frames. Because of increased information about window frames the consumers change their preferences over time. Hence, the distribution of the demand over the four types of window frames also changes over time, as can be seen in Figures 9.4 and 9.5 for new and renovated houses, respectively. The preferences change smoothly over time.

[2] This influence may be largely responsible for developments in the city of Amsterdam. Its policy was to stimulate the use of pinewood, resulting in an increase in its use from 1 to 52% in 2 years. Consequently, the use of hardwood decreased from 52 to 27%, aluminium from 30 to 21%, and pvc from 14 to 2% (Fraanje *et al.*, 1992).

[3] Stella II is a software package for developing dynamic models and performing dynamic simulation (Peterson and Richmond, 1994). Hannon and Ruth (1994) provide a good introduction to dynamic modelling using Stella II.

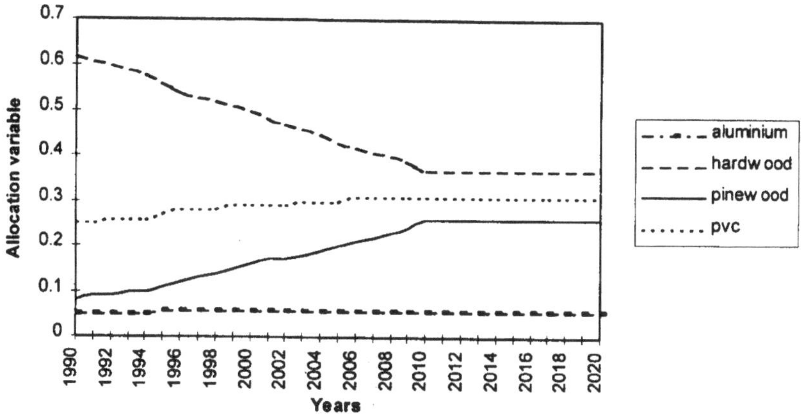

Figure 9.4. The allocation of the demand for window frames for new houses in Scenario 2.

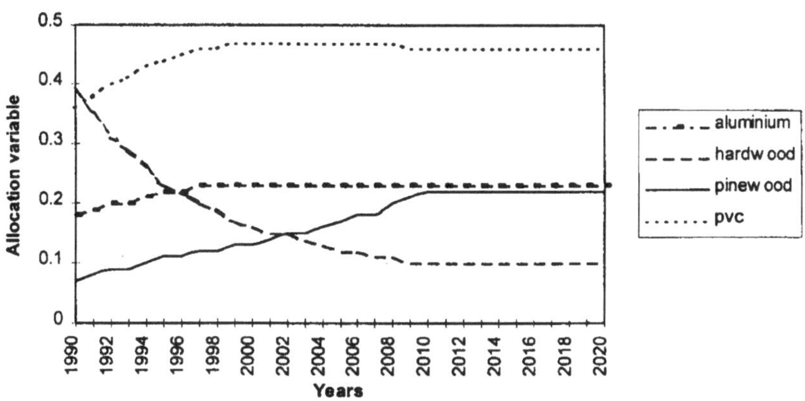

Figure 9.5. The allocation of the demand for window frames for renovated houses in Scenario 2.

The total demand is lower than in the reference scenario because the levy on the use of hardwood window frames increases the average price of window frames, so that a certain proportion of the houses are subjected to renovation of window frames.

Under the water pollution scenario (Scenario 3) the levy imposed on the use of aluminum window frames and the changes in preferences in favour of pinewood window frames result in a change in the distribution of the demand. The allocation of the demand for new window frames is similar to that of Scenario 2. Environmental repercussions are presented for four environmental performance indicators (Tables 9.4 to 9.7).

Table 9.4 shows that under Scenario 1 the depletion of raw materials is reduced because the purchase of window frames falls over time as the number of new and renovated houses drops. This reduction will even occur when the government does not impose any policies. Note that in 2010 acidification is somewhat higher than in 2000. This is due to a slightly higher demand for houses in 2010 (101,000 houses) than in 2000 (100,400 houses). Under all scenarios and for all performance indicators a similar effect operates.

Under Scenario 2 the use of hardwood is reduced while the recycling of hardwood waste is increased. The results under this scenario can be compared with those under Scenario 1 where no instruments are used. It is clear that depletion can be more than 50% over a time span of 20 years (compare 3279 and 1607).

Under Scenario 3 it is shown that when pollution is reduced by using less aluminum and pvc, the use of raw materials is also reduced. This reduction is smaller than when instruments are imposed on the use of hardwood, as under Scenario 2.

Table 9.4. Raw materials depletion.

Year	Raw materials depletion (in 10^6 depletion units)		
	Scenario 1	Scenario 2	Scenario 3
1990	6051	6051	6051
1995	4979 (100%)	3897 (78.2%)	4306 (86.5%)
2000	3256 (100%)	2009 (61.7%)	2192 (67.3%)
2010	3279 (100%)	1607 (49.0%)	1695 (51.7%)

In Table 9.5 the results under Scenario 3 are compared with the results under Scenarios 1 and 2, where no specific policies are used to reduce water pollution. The results are now presented over a longer period of time because water pollution not only occurs when the window frames are produced and used, but also after 30 years, when the window frames give rise to a waste flow. In the case of resource depletion no delayed effects will occur because the resources are only used at the production stage.

The amount of acidification in both the production and use stage is presented in Table 9.6. As a result of the decrease of total demand for window frames because of the reduction in the demand for houses, acidification decreases in the reference scenario. Under Scenario 2 total demand for window frames falls because of the reduction in the demand for houses, and because of the levy on hardwood window frames. Under Scenario 3 the endogenous reduction of the demand for window frames is caused by a levy on the use of aluminum window frames. The change in the distribution of the total demand also entails that under Scenarios 2 and 3 there is less acidification over time. Furthermore, acidification under Scenario 3 is less than under Scenario 2 since the use of the type of window frame that causes relatively much acidification is discouraged under Scenario 3.

Table 9.5. Water pollution.

Year	Water pollution (in 10^6 m^3 potentially polluted water)		
	Scenario 1	Scenario 2	Scenario 3
1990	483.4	483.4	483.4
2010	483.2 (100%)	469.8 (97.2%)	395.8 (81.9%)
2030	397.5 (100%)	359.7 (90.4%)	255.2 (64.2%)
2050	399.0 (100%)	367.7 (92.2%)	259.4 (63.4%)

The use of energy (not shown) shows the same pattern as the acidification, which makes sense because the decrease of the demand and the change in distribution of the demand over the four types of window frames have the same direct effects on energy use as on acidification. The use of energy is the lowest under Scenario 3 because then discouragement of the use of aluminum window frames is most effective.

Global warming potential is the highest for hardwood window frames. Under Scenario 2 the use of this window frame type is reduced which lessens the global warming potential more than in the other two scenarios.

Table 9.6. Acidification.

Year	Acidification (in 10^6 acidification units)		
	Scenario 1	Scenario 2	Scenario 3
1990	2023	2023	2023
1995	1704 (100%)	1645 (96.5%)	1382 (81.1%)
2000	1198 (100%)	1006 (84.0%)	814 (67.9%)
2010	1204 (100%)	990 (82.2%)	789 (65.5%)

Total waste material, presented in Table 9.7, arises from the production and waste treatment of window frames. Not only at the production stage, but also thirty years after production materials in window frames become free materials (waste) flows. Such a delayed effect also occurs with respect to water pollution associated with discarded window frames. Note that other environmental indicators, such as depletion, do not suffer such a delayed effect. For hardwood and pinewood window frames the generation of waste at the production stage is lower than the waste after consumption. The total waste generated is higher for pvc and aluminum than for hardwood and pinewood so that a reduction in the use of pvc and aluminum affects the generation of waste more than under the second scenario. In Scenario 3 the use of aluminum and pvc is reduced so that under this scenario the generation of waste is more reduced than under Scenario 2.

Table 9.8 gives an overview of the effects of separate instruments in terms of the six environmental indicators. The policies are ranked from 1 to 5, with 1 as the most effective policy in terms of the respective environmental indicator. Details about the analysis underlying the ranking, based on a simple multi-criteria analysis, are presented in Kandelaars and Van den Bergh (1996b). This can be considered to be a type of sensitivity analysis because it generates results for a variety of policy formulations. Other types of sensitivity analysis, using Monte-Carlo type of approaches, for instance, by randomizing economic and environmental effects, may be performed. Also sensitivity of rankings with respect to weights can be done. This would, however, be less useful in the present context as it takes us away from the materials-economy modelling which is focused on here. For further details about sensitivity analysis in the context of multi-criteria analysis (Janssen, 1992).

Table 9.7. Waste.

Year	Waste (in 10^6 kilograms)		
	Scenario 1	Scenario 2	Scenario 3
1990	183.3	183.3	183.3
2000	144.7 (100%)	129.0 (89.1%)	130.7 (90.3%)
2010	141.7 (100%)	124.5 (87.9%)	120.5 (85.0%)
2030	101.7 (100%)	82.4 (81.0%)	71.7 (70.5%)
2050	102.0 (100%)	84.3 (82.6%)	72.1 (70.7%)

Charges on window frames appear to have the best overall effects. This is partly due to the reduction in the demand for window frames when the average price is sufficiently high. The levies on dumping and use of new materials have different effects depending on the indicator.

Table 9.8. Ranking of the performance of separate instruments per environmental indicator.

Environmental indicator	Ranking				
	Levy on dumping of hardwood	Levy on hard-wood window frames	Levy on aluminium and pvc dumping	Levy on use of new aluminium and pvc	Levy on aluminium window frames
Raw materials depletion	3	1	5	4	2
Water pollution	5	3	1	4	2
Acidification	5	2	4	3	1
Energy use	5	2	3	4	1
Global warming	3	1	4	5	2
Waste	3	2	5	4	1

A multi-criteria analysis can be used to evaluate these environmental indicators. When equal weights are used for each indicator the conclusion is that a charge on aluminum window frames is the best single instrument, followed by a charge on hardwood window frames, a levy on aluminum and pvc dumping, and in a shared fourth place a levy on dumping of hardwood and a levy on the use of new aluminum and new pvc.

The performance of the M-P chain under the different scenarios in terms of economic indicators is based on the average costs per window frame for the consumers, and on the revenue of the levies paid by consumers (Tables 9.9 and 9.10).

In Table 9.9 the average cost per window frame changes under Scenarios 2 and 3. The different types of window frames have different prices which makes it more convenient to look at the average cost. Under Scenario 2 the average cost first increases because of the levy on hardwood window frames. After this initial increase the average cost decreases because more pinewood window frames are bought and these are cheaper than the other three types of window frames. Under Scenario 3 the average cost decreases gradually. The increase in the average cost due to the levy on aluminum window frames is less than the decrease in the average cost due to the shift towards less expensive window frames.

The total amount of levies paid decreases over time because of a reduction of the general demand for window frames and a reduction of the demand for specific types of window frames on which levies are imposed, as shown in Table 9.10. Under Scenarios 2 and 3 the number of window frames purchased in 2010 is higher than in 2000, but the total amount of levies paid by consumers decreases. This is due to the reduction of the demand for specific types of window frames, namely, hardwood window frames in Scenario 2 and aluminum ones in Scenario 3. All in all, significant changes over time as well as differences between scenarios and instruments are indicated by these simulation results.

Table 9.9. The average cost that consumers pay per window frame.

Year	Average cost per window frame for consumers (in 10^3 Dutch Guilders)		
	Scenario 1	Scenario 2	Scenario 3
1990	4.84	4.84	4.84
1995	4.84	4.98	4.82
2000	4.84	4.86	4.75
2010	4.84	4.70	4.64

Table 9.10. Revenue for the government from levies on products paid by consumers.

Year	Revenue for government from product levies paid by consumers (in 10^6 Dutch Guilders)	
	Scenario 2	Scenario 3
1990	0	0
1995	57.4	14.1
2000	28.3	10.0
2010	20.9	10.1

9.7. Conclusions

An M-P chain analysis was performed for window frames. These can be produced from different materials each of which can be partly recycled. Their use is widespread and has a number of indirect environmental impacts. Six environmental and two economic indicators were used to evaluate three different policy scenarios and a set of policy instruments. The policy scenarios each focus on specific environmental problems, i.e. the depletion of raw materials and water pollution. The model is based on a vintage structure of the stock of window frames in use. Each year window frames are demanded for implementation in new and in existing houses (replacement). The impact of levies on window frames is twofold: a reduction of the total demand for window frames; and, a change in the distribution of demand over the four types of window frames.

The policy package focusing on water pollution gives better results in terms of the indicators for water pollution, acidification, energy use and waste than the policy oriented towards depletion of raw materials. Likewise, for global warming and raw materials depletion the focus on the depletion of raw materials thus seems more effective. As separate instruments, the levies on window frames are the most effective in terms of all environmental indicators except water pollution. To reduce water pollution levies on the dumping of aluminum and pvc turn out to be the most effective. The average cost per window frames for consumers declines over time because of the change in prices and preferences. The revenues for the government from the levies on the window frames decrease over time because of a decrease in the total demand and a decrease in the demand for the specific window frame charged.

Studies based on M-P chain models can increase our understanding and possibly long-run prediction of complex dynamic interactions between preferences, policies, technology and materials demand and supply. In particular, they offer more support for long-run environmental policy than tools for pure material flow analysis, such as MFA/SFA. Further work is needed to develop the modelling of technical processes and economic decisions made by producers, including those associated with recycling of products and materials. Other interesting future research can consider the dynamic and evolutionary aspects of material and product flows. This research can be linked with recent dynamic MFA and LCA studies (Moll, 1993; Gilbert and Feenstra, 1994; Berg, 1994). An evolutionary approach would allow better care to be taken of issues like uncertainty; co-evolution of policy, preferences and technology; and interaction between different scales of analysis (e.g. Dosi *et al.*, 1988; Faber and Proops, 1990). It should be immediately noted, however, that such an approach is difficult to implement based on empirical data only. Assumptions are inevitable regarding the precise character of various dynamic processes. Furthermore, this approach can, when formally tackled, give rise to complex model formulations (e.g. Clark *et al.*, 1995). Very specific, descriptive aspects of products like design, de-composability and packaging can of course hardly be taken into account in formal dynamic model formulations. Instead, descriptive or evaluative expert studies seem essential complements to the formal modelling discussed here.

Appendix 9. The equations of the model with explanation

In this appendix the most important equations of the dynamic simulation model are listed and some are provided with a short explication, including the specific choices made with each scenario.

The following legend explains the abbreviations used. For the abbreviations for materials or products only the abbreviations for aluminium is given, because those for hardwood, pinewood and pvc can be derived from these.
new = the demand for new houses
renov = the demand for renovated houses
demwfnew = demand for window frames for new houses
demwfren = demand for window frames for renovated houses
averpwf = average price for a window frame
renov = renovated houses
levywfalu = levy on window frames made of aluminium
levyalu = levy on new aluminium
levydumpalu = levy on dumping of aluminium
pwfalu = price of the window frame made of aluminium
pnewalu = price of aluminium
paludump = price of aluminium dumping
precalu = price of recycled aluminium
pwasterecalu = price of waste of aluminium for recycling
wfalu_in = window frames made of aluminium in the economy or the system
disalu = discount factor for aluminium when recycled aluminium is used for production
disaluwaste = discount factor for aluminium waste when this waste is used for recycling
acidif_a = acidification caused by the use of aluminium window frames
energy_a = energy use caused by the use of aluminium window frames
envwaste_a = environmental waste caused by the use of aluminium window frames
globwarm_a = global warming caused by the use of aluminium window frames
raw_mat_a = raw materials depletion caused by the use of aluminium window frames
waterpol_a = water pollution caused by the use of aluminium window frames

Scenario 1

Demand

1) demwfnew = new*21.71/6
The demand for new window frames is the number of new houses. One house needs 21.71 m^2 window frames and the functional unit is 6 m^2 (Hoefnagels *et al.*, 1992).
2) new = GRAPH(time)
(1990, 109), (1991, 109), (1992, 109), (1993, 109), (1994, 109), (1995, 84.6), (1996, 84.6), (1997, 84.6), (1998, 84.6), (1999, 84.6), (2000, 44.4), (2001, 44.4), (2002, 44.4), (2003, 44.4), (2004, 44.4), (2005, 45.0), (2006, 45.0), (2007, 45.0), (2008, 45.0), (2009, 45.0)
The number of new houses changes over time.
3) renov = GRAPH(time)
(1990, 65.0), (1991, 65.0), (1992, 65.0), (1993, 65.0), (1994, 62.0), (1995, 61.0), (1996, 58.0), (1997, 58.0), (1998, 57.0), (1999, 57.0), (2000, 56.0), (2001, 56.0), (2002, 56.0), (2003, 56.0), (2004, 56.0), (2005, 56.0), (2006, 56.0), (2007, 56.0), (2008, 56.0), (2009, 56.0)
The number of renovated houses changes over time.
4) demwfren = if averpwf<4600 then renov*21.71/6 else renov*change*21.71/6
5) change= if time>1999 then 0.75 else 1-(time-1994)*0.05
The demand for window frames for renovated houses equals the number of renovated houses if the average price of window frames is less than 4600 guilders, otherwise this demand is a decreasing percentage of the number of renovated houses.

Prices (in Dfl)

6) pwfalu = 16772*6/21.71 + levywfalu

7) pwfhard = 18887*6/21.71 + levywfhard

8) pwfpine = 14504*6/21.71

9) pwfpvc = 16014*6/21.71

The price of one aluminium, hardwood, pinewood and pvc window frame (Hendrix and Martens, 1990).

10) levywfalu = levywfhard = 0

The levies on window frames are zero.

11) averpwf = (pwfalu+pwfhard+pwfpine+pwfpvc)/4

The average price of a window frame.

12) pnewalu = 1 + levyalu

13) pnewhard = 1 + levyhard

14) pnewpine = 1 + levypine

15) pnewpvc = 1 + levypvc

The prices of new aluminium, hardwood, pinewood and pvc.

16) levyalu = levyhard = levypine = levypvc = 0

The levies on new materials are zero.

17) precalu = prechard = precpine = precpvc = 1

The prices of recycled aluminium, hardwood, pinewood and pvc.

18) pwasterecalu = pwasterechard = pwasterecpine = pwasterecpvc = 1

The prices of waste aluminium, hardwood, pinewood and pvc for recycling.

19) paludump = 1 + levyaludump

20) pharddump = 1 + levyharddump

21) ppinedump = 1 + levypinedump

22) ppvcdump = 1 + levypvcdump

The prices of dumping aluminium, hardwood, pinewood and pvc.

23) levyaludump= levyharddump = levypinedump = levypvcdump = 0

The levies on dumping materials are zero.

Weights of window frames (in kilograms)

24) weight of one aluminium window frame = 15.77

25) weight of one hardwood window frame = 63.73

26) weight of one pvc window frame = 15.16

27) weight of one pinewood window frame = 45.68

Derived from Hoefnagels *et al.* (1992).

The environmental burden caused by aluminium window frames[4]

28) acidif_a = wfalu_in*(5520+2270)*(1-0.5*disalu)*(1-0.3*disaluwaste)

The acidification as a result of the use of aluminium window frames is per window frame 5520+7720 ZE (ZE=acidification unit) when no recycling takes place. When recycled materials are used in the production stage of window frames the acidification decreases by 50%. Also when the materials are recycled again after use the acidification now decreases by 30%. The same applies for energy, global warming and energy use.

29) energy_a = wfalu_in*(7.720+31.600)*(1-0.1*disalu)*(1-0.1*disaluwaste)

30) envwaste_a = (wfalu_in*(158+43.7)+(wfaluwaste*73.6))*(1-0.35*disalu)*(1-.25*disaluwaste)

Window frames that are dumped are an important part of the environmental waste which is the

---

[4] The data for the environmental burden of the four types of window frames is derived from Hoefnagels *et al.* (1992).

second term on the right side. The environmental waste occurs in the products and use phases (158+43.7) and in the dumping phase which has a lag of 30 years (73.6) whereby this amount is calculated for the dumped window frames in that year.

31) globwarm_a = wfalu_in*(645+1710)*(1-0.1*disalu)*(1-0.1*disaluwaste)
32) raw_mat_a = wfalu_in*(397+55.2)*(1-0.05*disalu)*(1-0.03*disaluwaste)
33) waterpol_a = (wfalu_in*(1650+11.7)+(wfaluwaste*1030))*(1-0.55*disaluwaste)*(1-0.2*disalu)

The environmental burden caused by hardwood window frames

34) acidi_h = wfhard_in*(1480+1700)*(1-0.03*dishardwaste)
35) energy_h = wfhard_in*(1.78+19.8)+(1-0.005*dishardwaste)
36) env_waste_h = (wfhard_in*(51.7+53)+(wfhardwaste*129))*(1-0.42*dishardwaste)
37) glob_warm_h = wfhard_in*(1450+1440)*(1-0.08*dishardwaste)
38) raw_mat_h = wfhard_in*(13100+3420)*(1-0.12*dishardwaste)
39) waterpol_h = (wfhard_in*(9.41+31.4)+wfhardwaste*471)*(1-0.25*dishardwaste)

Recycled hardwood cannot be used for hardwood window frames.

The environmental burden caused by pinewood window frames

40) acidi_pi = wfpine_in*(1170+1190)*(1-0.03*dispinewaste)
41) energy_pi = wfpine_in*(2.04+15.8)*(1-0.01*dispinewaste)
42) env_wastepi = (wfpine_in*(57.3+32.5)+(wfpinewaste*143))*(1-0.48*dispinewaste)
43) globwarm_pi = wfpine_in*(297+926)*(1-0.02*dispinewaste)
44) raw_mat_pi = wfpine_in*(881+347)+0*dispinewaste
45) waterpol_pi = (wfpine_in*(16.1+22.5)+wfpinewaste*521)*(1-0.27*dispinewaste))

Recycled pinewood cannot be used for pinewood window frames.

The environmental burden caused by pvc window frames

46) acidi_p = wfpvc_in*(891+1060)*(1-0.02*dispvc)*(1-0.03*dispvcwaste)
47) energy_p = wfpvc_in*(3.54+24.9)*(1-0.01*dispvc)*(1-0.02*dispvcwaste)
48) env_wastep = (wfpvc_in*(171+69.1)+(wfpvcwaste*125))*(1-0.21*dispvc)*
(1-.21*dispvcwaste)
49) globwarm_p = wfpvc_in*(322+1350)*(1-0.01*dispvc)*(1-0.03*dispvcwaste)
50) raw_mat_p = wfpvc_in*(1630+436)*(1-.25*dispvc)*(1-0.25*dispvcwaste)
51) waterpol_p = (wfpvc_in*(192+61.6)+(wfpvcwaste*1280))*(1-0.18*dispvcwaste)*
(1-.18*dispvc)

Discount factors

52) disalu = if pnewalu+levyalu>precalu then 0.9 else 0
53) disaluwaste = if paludump+levyaludump>pwasterecalu then 0.9 else 0
54) dishard = if pnewhard>prechard then 0.8 else 0
55) dishardwaste = if pharddump+levyharddump>pwasterechard then 0.75 else 0
56) dispine = 0
57) dispinewaste = if ppinedump+levypine>pwasterecpine then 0.75 else 0
58) dispvc = if pnewpvc+levypvc>precpvc then dispvcchange else 0
59) dispvcwaste = if ppvcdump+levypvcdump>pwasterecpvc then dispvcchange else 0
60) dispvcchange = max(0.04*(time-1990),0.8)

The discount factors for new materials depend on the prices of new materials, the levies on new materials and the prices of recycled materials. The discount factors for waste material depend on the prices of dumping materials, the levies on dumping and the prices of materials to be recycled.

The environmental burden caused by the sum of all window frames

61) acidif = acidif_a+acidi_p+acidi_pi+acidi_h
62) energy = energy_a+energy_p+energy_pi+energy_h
63) env_waste = envwaste_a+env_wastep+env_wastepi+env_waste_h

64) globwarm = globwarm_a+globwarm_p+globwarm_pi+glob_warm_h
65) raw_mat = raw_mat_a+raw_mat_p+raw_mat_pi+raw_mat_h
66) waterpol = waterpol_a+waterpol_h+waterpol_p+waterpol_pi

The sum of the environmental impacts of the four window frames is the total environmental impact.

Scenario 2

Prices

67) levywfhard = 250
68) levydumphard = if time > 1994 then 1 else 0

The levy on the dumping of hardwood means that recycling of hardwood becomes cheaper than dumping. The highest technically possible amount of hardwood will be recycled. This is set to be 75% (dishardwaste=0.75 instead of 0 in Scenarios 1 and 3).

Scenario 3

Prices

69) levywfalu = 250
70) levyalu = 1
71) levypvc = 1
72) levydumpalu = 1
73) levydumppvc = 1

The levies on the use of new aluminium and the dumping of aluminium mean that 90% of the aluminium is recycled when the levy is set (disalu = 0.9 and disaluwaste = 0.9). The levies on the use of new pvc and the dumping of pvc mean that pvc is recycled more. When the levy is imposed the percentage of recycling is 20% in 1995, going up to 80% in 2010. (dispvc and dispvcwaste are increasing over time; 1990-0%, 1995-20%, 2000-40%, 2005- 60%, 2010-80%)

CHAPTER 10

MATERIAL FLOWS IN AN APPLIED GENERAL EQUILIBRIUM MODEL[1]

10.1. Introduction

This chapter tries to set a first, empirical step in filling the gap between physical and economic models, by combining a material flow model and a disaggregated applied general equilibrium (AGE) model. The goal of this model is to empirically assess the economy-wide and environmental effects of environmental policies focusing on particular materials.

A material flow model describes the physical flows of materials through the various sectors of an economy, so that material balance is satisfied for each material in each sector in the model (see Section 4.3 in Chapter 4). To change the use of materials in an economy a material policy may be imposed on specific materials in certain sectors. With a disaggregated AGE model the sectoral and distributional effects of policies can be analysed for various production sectors and household groups. Moreover, the effects of various policies on trade and employment may be examined.

The general equilibrium described in Chapter 7 includes material balance conditions and material flows between the economic activities. The equilibrium model is theoretical and it only describes an aggregate M-P chain and abstract sectors. The goal of the model was to assess the optimal policies. In this chapter an empirical equilibrium model for the Netherlands is used in which various production sectors and household groups are included. The AGE model is a model in monetary terms. Therefore, it needs to be combined with a material flow model in physical terms to be capable of linking monetary and physical flows.

A number of scenarios are designed to study the effects of a material policy on metals, in particular zinc and lead. These metals have been selected because of the environmental and health risks they may create (Gorter, 1994; Annema *et al.*, 1995). The data on the zinc and lead flows in the material flow model are used together with the monetary flows to calculate the height of the levies that are imposed on the various sectors in the AGE model for the Netherlands. Thus, the material flow model is used to determine policies in the AGE model. There is no endogenous feedback from the AGE model into the material flow model. The results of the AGE model are exogenously combined with the material flow model in order to provide the effects of the policy on the use of materials.

The organization of this chapter is as follows. Section 10.2 gives a short

[1] This chapter is based on Kandelaars and Dellink (1999).

description of the AGE model. Section 10.3 describes the material flow model and the integration of this model into the AGE model. Section 10.4 discusses various material and product policies that will be studied in the scenarios. Results of the scenario analysis are given in Section 10.5. The last section draws conclusions and presents suggestions for further research.

10.2. Description of AGE Models and the Taxinc-Model

The AGE model used in this chapter is the Taxinc-model.[2] The model was originally developed to analyse the effects of changes in the tax structure in the Netherlands (Keller, 1980; Cornielje, 1990; CBS, 1991) and has recently been adapted and applied to study the effects of energy levies (Dellink and Jansen, 1995). The present section briefly discusses AGE models in general and of these the Taxinc-model in particular.

Applied general equilibrium models

Applied general equilibrium (AGE) models are general equilibrium models that are empirically calibrated. The models consist of a set of aggregated economic agents, e.g. households and production sectors. These agents demand and supply 'goods', which are either consumption goods or production factors. The demand and supply of these agents are matched on markets, resulting in prices for all goods. The behaviour of agents is assumed to be rational, which means that they optimize utility (consumers) or profit (producers) subject to a set of constraints, e.g. a budget constraint for consumers and a production function for producers. The AGE model achieves an equilibrium on each market when demand equals supply, with prices to match.

In AGE models, not all economic agents are modelled separately, but they are aggregated into different categories. In other words, sectors or groups are assumed to operate as individual optimizing agents. The behaviour of agents is simplified: for example, a Cobb-Douglas production function is used, and some factors may be exogenously determined.

In contrast with macroeconomic models (see Section 5.5), general equilibrium models are not meant to forecast aggregate economic indicators (e.g. the balance of payments, unemployment and inflation) or the business cycle. AGE models have been developed to answer questions on a more disaggregated level than macroeconomic models. The questions may be about the interactions between taxation, economic development, welfare distribution, allocation of goods and trade.

The data needed for an AGE model are based on a (social) accounting matrix, in which input-output tables, national accounts, consumption and, if relevant, taxation are integrated, and price data (Cornielje, 1990). An AGE model is calibrated using these data. A static AGE model is calibrated for a certain year.

---

[2] 'Taxinc' stands for tax incidence.

Taxinc-model

The AGE model used in this study is the Taxinc-model (Keller, 1980; Cornielje, 1990). The economic agents in the Taxinc-model are production sectors (producers), household groups (consumers), and a fiscal agent. The demand and supply equations of production sectors and household groups are derived from standard neo-classical economic theory, implying that all agents behave rationally, firms maximize their profits given prices and capacity, and households their utility given prices and transfers. Market prices are determined so that all markets clear. The price of a product paid by buyers equals the market price plus taxes levied on the demand side of the market. Suppliers face a price equal to the market price minus supply-side taxes. The fiscal agent collects the taxes and redistributes them lump-sum to the household groups, e.g. as social security benefits. The structural characteristics of the model, such as household preferences, production structure, production capacity and technology are exogenous.

The model includes 61 production sectors and 44 household groups allowing the analysis of sectoral and distributional consequences of policies.[3] Each production sector is assumed to produce a single, unique good, so individual firms within that sector are assumed to be identical. Each household group represents individual households with identical marginal shares, i.e. every household within the sector is assumed to spend extra income in exactly the same way. Households own the production factors labour and capital. The utility functions of the household groups and the production function of the production sectors are of a nested CES-type (Constant elasticity of substitution) (Keller, 1980; Cornielje, 1990).[4] The public sector is divided into two parts: (i) a household group that consumes public goods, called the public sector; and, (ii) a production sector, called public services, that produces public goods. Demand from foreign firms and households (exports) are specified as the demand by a single representative utility-maximizing household group called 'Rest of the world' (Woodland, 1980). The 'small open economy assumption' is used which implies that world market prices are exogenous. This assumption results in a strong reaction by the foreign sector to changes in prices. The Taxinc-model includes the 'Armington assumption' that distinguishes import and export of a good as two different goods. This heterogeneity between imported and exported goods means those two categories are imperfectly substitutable. In the Taxinc-model, imports are themselves differentiated into competitive imports, for natural gas, crude petroleum and other goods and services. Non-competitive imports, for which there are no domestic substitutes. These non-competitive imports go directly from the 'Rest of the world' to the other household groups (for further details, see CBS, 1991).

[3] This aggregation in production sectors and household groups is based on CBS (1991).

[4] An example of a nested utility function is as follows: aggregate consumption is divided into necessities and luxuries, the necessities are divided among food and clothes, food is divided into vegetables, milk and bread (Keller, 1980). Thus, a nested function has a tree-structure. A CES production function with output (Q) and 3 inputs (K=capital, L=labour, R=resources) has the following structure $Q=A(\alpha K^{-\theta}+\beta L^{-\theta}+\gamma R^{-\theta})^{-1/\theta}$ with $A,\alpha,\beta,\gamma>0$, $\alpha+\beta+\gamma=1$ and $-1<\theta<>0$.

The Taxinc-model is a static equilibrium model. For the calibration of the Taxinc-model the base year is 1988, including the current policies in the Netherlands in that year. An exogenous change in the tax structure results in a new equilibrium that is determined by the initial equilibrium and the marginal reactions of producers and households. Therefore, with the Taxinc-model a comparative static analysis may be performed, which allows the comparison of two equilibria. The adjustment process and the (time) path from the initial to the new equilibrium, and intertemporal effects are not considered. The tax impulse causes relative prices to shift, inducing a change in the demand for and supply of all goods and services. These demand and supply changes in effect cause a change in market prices and hence a change in relative prices. This results in a new equilibrium, where all markets clear again. In this new equilibrium, relative prices and demand for and supply of all goods and services may differ from the initial equilibrium.

10.3. Integrating the Material Flow Model 'Flux' with the Taxinc-Model

The Taxinc-model that is described in Section 10.2 will now be combined with a material flow model 'Flux'. First, Flux will be examined and then the integration of Flux and the Taxinc-model is described.

The material flow model 'Flux'

A material flow analysis (MFA) describes the flow of one or more specific materials in a geographic area during a certain period of time (see Section 4.3 of Chapter 4). The basis of an MFA is a database on stocks and flows of materials. The material flow model 'Flux' consists of a database that describes the physical flows of materials and products through the Dutch economy in the year 1990 (Boelens and Olsthoorn, 1998). It is an input-output model in physical units, that describes the input and the output of material flows of various sectors so that material balance conditions hold for each sector of the economy (see for physical input-output models, Section 4.4 of Chapter 4). The material flow model consists of domestic and foreign economic sectors and an environmental 'sector' (that is divided into various compartments such as atmosphere, soil and water). The inflow of materials into the domestic economic system originates both from the foreign economic sector (i.e. imports) and the environmental sector (i.e. extraction of new materials).

In addition to a database, Flux may be used for (linear) modelling to answer questions like: 'What if the recycling of product X increases by 10%?'. A unique feature of Flux compared with other material flow models is that an incomplete database can be completed by different 'balancing procedures' in order to balance the inputs and outputs in a sector.

Combining the material flow model 'Flux' with the AGE model 'Taxinc'

For the combination of Flux and Taxinc, the domestic economic sectors of Flux are connected to the production sectors in the Taxinc-model. Both the sectoral division in Flux and Taxinc is based on the data of CBS, which facilitates the connection of the sectors in both models.

The data on the material flows in Flux are used to implement material policies in the Taxinc-model. For this implementation, one or more materials are chosen. The physical data on the material flows between economic sectors (including extraction) are connected to the data in monetary units in the Taxinc-model. A levy on an economic sector is determined by the use of materials in kilograms (physical data) and the output in guilders (monetary data) of that sector.

An example is a levy on the imports of aluminium. This policy may be imposed on the various sectors depending on the use of this material (in kilograms) and on their input or output (in monetary terms) in that sector (see Section 10.4).

After the levy is imposed the Taxinc-model calculates a new equilibrium. The results of the Taxinc-model are imported into Flux to assess the effects of a policy on the material flows. Thus, the combination of Flux and the Taxinc-model allows for the construction of various policy-scenarios to examine the economic and physical (environmental) effects of such policies.

Within the current model set-up it is assumed that the 'material intensity', which is defined as the material input (in kilograms) per guilder of output, of the production sectors has not changed between 1988, the year for which the Taxinc-model is calibrated, and 1990, the base-year of Flux. Furthermore, it is assumed that the material intensity does not change when a levy is imposed. This allows the changes in output levels (in monetary units) of production sectors to be translated into changes in the output levels in physical units and in the use of materials. Clearly, this effect on materials will differ from the 'true' effects, *inter alia* because changes in the material contents of inputs cannot be accounted for, and because substitution within a sector is not taken into account (for substitution, see Section 3.4.2 of Chapter 3). However, substitution of one material by another is considered if the supplying production sector is different for both materials.

10.4. Material and Product Policies

A material policy may be imposed to reduce the use of definite materials and products (see Section 3.5.1 of Chapter 3). Such policies generate revenues for the government. The policies used in this chapter are 'regulatory policies', and their revenues are redistributed to the taxpayers. Therefore, a scenario for a policy implementation consists of two parts: (1) the material or product policy; and, (2) the destination of the revenues generated by the policy. In practice, policy makers need to make choices about both of these issues. In this section the policies will be described in general terms, while in Section 10.5 the specific instruments in each scenario are discussed.

The scenario analyses are not performed for the purpose of forecasting, but for illustrating the tax mechanisms in an empirical model in which material flows are combined with an AGE model. Although the Taxinc-model, as all AGE models, has its weaknesses, it is an appropriate tool to give a good insight into the disaggregated distributional and sectoral effects of policies.

The first part of each scenario concerns material and product policy and involves choosing which particular material flows will be levied and the height of the levy.

The height of the levy for the production sectors and household groups depends on their use of certain materials and for the production sectors also on their output in monetary units. For material policies, the levy depends on the 'material intensity' of the production sector which is the material input (in kilograms, obtained from the material flow model) divided by the output (in monetary units, obtained from the AGE model). This material intensity connects the material flows with the monetary flows. The material intensity of a sector determines the relative tax weight. Thus, the absolute tax payments per guilder of output by a sector equals the relative tax weight multiplied by the overall tax rate. The overall tax rate of the levy is arbitrarily chosen such that the total tax revenue will be 0.1% of national income.

The second part of the scenario is to choose the destination of the revenues generated by the material policy. The levies imposed on physical flows are formulated as regulatory policies and the revenues of these are redistributed to the taxpayers. This stands in contrast to a revenue-raising tax that is used to generate public income. The revenues are redistributed by reducing the labour taxes paid by the production sectors, which may be interpreted as a reduction in the contributions of employers to social security.

The material levy and the reduction of the labour tax result in a shift in the tax burden from labour to material use. The current tax system discourages work and investment, and it encourages pollution and material use. Therefore, a tax shift from labour to material use is attractive, because it taxes more what we want less of, i.e. material use, and less what we want more of, i.e. labour (Hamond *et al.*, 1997).[5] In model terms, this is achieved by introducing a subsidy on the demand for labour.

This regulatory levy can be seen as a 'green tax reform'. However, though the explicit objective is to reduce the use of specific materials, stimulating employment may be an implicit goal. If both goals are satisfied, i.e. a positive effect on employment and environment is achieved, then a 'double dividend' is reaped. In the discussion of the results in Section 10.5, the possibility of a double dividend will be addressed.

Some characteristics are common for all scenarios. Every scenario consists of a sector-specific levy on the use of certain materials. Only the domestic demand for materials or products is levied; hence, exports are exempt from levies. Imports are levied in the same way as domestic goods and services. Implicitly, the material contents of the imported products are assumed to be equal to those of domestic substitutes.[6]

The regulatory levy will not only affect directly the sectors that are subjected to the levy, but indirectly also other production sectors and household groups. The levied sectors will change their demand for inputs as their costs rise. This leads to a

---

[5] Other possibilities to redistribute the revenues to the taxpayers are to subsidize the use of an environmentally less-damaging input, introduce a subsidy on R&D, or a lump-sum transfer to the production sectors or households.

[6] As imports in the material flow model are not distinguished by origin, these imports cannot be linked to the imports in the AGE model. Therefore, it is necessary to make the assumption that the material intensity of foreign production is equal to that of domestic production.

'backward effect' through the economy: the suppliers of these inputs are confronted with a lower demand, and consequently their production and their demand for inputs decreases. Next, the suppliers of these inputs see their demand lowered and will react to that, and so on. Furthermore, there is a 'forward effect' through the economy: as the demand for some goods and services decreases, the price for these goods and services will also decrease, which affects the buyers of these goods or services. These will also react to the changes in relative prices, which in turn triggers a reaction of other agents. In total, all relative prices will change, and consequently all production sectors and household groups are affected by the levy.

Figure 10.1 illustrates graphically the stages in production and consumption where material or product policies may be imposed. Three steps are identified. In the first production stage, raw materials are converted into (intermediate) products. Between the first and final production stages there may be several intermediate stages of production. All production sectors require inputs from, and supply products to, other economic sectors.

Moreover, the production stages require inputs which include 'inflows' of materials (extraction of raw materials and imports). All stages have 'outflows' of materials to the environment and to foreign economies (exports). An example of the three stages is the following: in the first stage the basic metal industry produces zinc plates; in the intermediate stage the metal product industry produces rain gutters; and, in the final production stage the construction sector fixes rain gutters to new houses.

A regulatory levy on primary material input

A regulatory levy on primary material input is imposed on the inflow of materials into the Dutch economy, either from imports from foreign economic sectors or extraction from the environment (see arrows labelled 1 in Figure 10.1). This means that the production sectors are levied according to the material intensity of the sector. The intermediate flows of materials between sectors are not levied, to avoid double-counting. In other words, materials are taxed at the moment they enter the domestic economy. Hence, the material intensities that are calculated from the inflows do not reflect the actual material contents of the produced goods and services in the sector. They only reflect the materials added during the production stage and not, for instance, the material contents of inputs.

A regulatory levy on materials throughput

With a regulatory levy on primary material input the first stage production sectors will be severely affected. Therefore, an alternative material policy may be imposed on the throughput of materials in the economy. This is simulated by levying the inflow of materials into all sectors, except for first stage production sectors (see arrows labelled 2 in Figure 10.1). The total material flow on which the policy is imposed is smaller than in the base scenario. This is compensated by a higher overall tax rate, such that the total tax revenue is equal to that of the base scenario. This scenario does not directly affect the first stage production sectors. However, the levy may influence these sectors indirectly: for example, by causing a reduced demand for metals by the other sectors.[7]

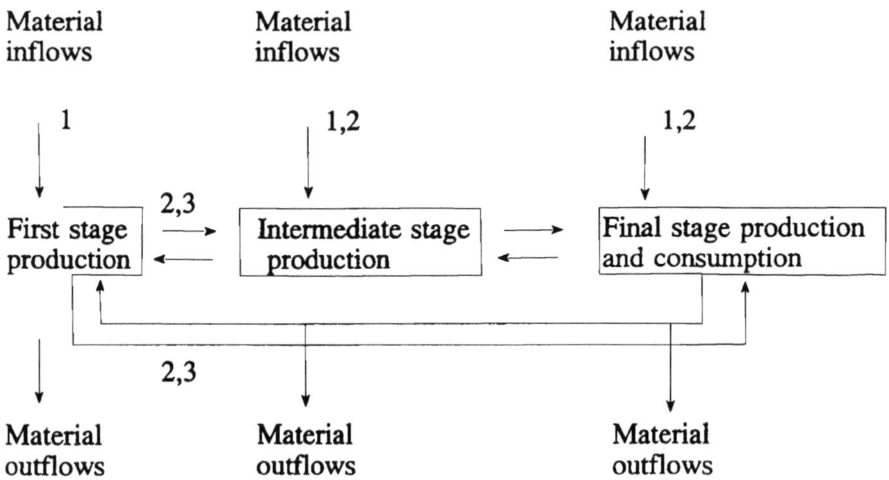

Figure 10.1. Stages in production and consumption where material policies may be imposed.
Note: Arrows indicate material and product flows; numbers are explained in the text.

A regulatory levy on products
The product policy is directed at products that contain certain materials. The consumers of these products, that may be either production sectors or household groups, are subjected to the levy (see arrows labelled 3 in Figure 10.1). The goal of this policy is to encourage consumers to switch to other products with a lower metal content. The main difference with the regulatory material levies is that the products are levied, not their material contents. In this scenario the relative tax weights are equal across all sectors. The total tax paid by a sector depends on the value of the products bought from the first stage production sectors.

In the scenario analysis the material policies will be simulated for two specific heavy metals: zinc and lead. As indicated before, in the case of a material policy the height of the levy depends on the material intensity of the sector. Table 10.1 presents the zinc and lead intensities for selected production sectors. This metal intensity is the use of that metal per unit of output, measured in kilogram/guilders. For illustrative purposes, the share in the total use of a particular metal is presented (measured as a percentage of the use in kilograms).

Table 10.1 shows that the share of the basic metal industry in the total inflow of

[7] Unfortunately, in this version of the model it was impossible to include a link between the throughput of materials between the sectors and the source of this throughput (the first stage production sectors).

zinc is enormous (more than 80% of the total zinc use). Other large users of zinc are the basic chemical industry, metal products manufacturing and trade (through imports). The lead intensity of the basic metal industry is much lower. Apart from these production sectors, the construction sector has a high lead intensity.

At first glance, the large zinc and lead intensity of the other sectors may seem surprising. However, these sectors encompass the waste treatment and processing firms that account for a large use of metals (Boelens and Olsthoorn, 1998).

It should be noted that a high (low) metal content of the inputs of a production sector does not necessarily imply a high (low) metal intensity of the goods and services produced in the sector. The reported intensities are based on the place where the metals enter the economic process. For example, a final stage production sector may produce goods and services with a high metal content, but add a little more metal in the production process itself (small inflow). The materials included in the economic inputs are already accounted for in previous production stages.

Table 10.1. Metal intensities of selected production sectors.

Production sector	Zinc intensity (kg/guilder)	Share of zinc use (%)	Lead intensity (kg/guilder)	Share of lead use (%)
Agriculture	0.01	0.2	0.00	0.0
Grain mills	0.06	0.3	0.00	0.0
Petroleum refineries	0.01	0.1	0.02	0.2
Basic chemical industry	0.44	6.3	0.31	6.5
Basic metal industry	12.30	81.4	4.78	46.0
Fabricated metal products	0.42	3.7	1.08	13.8
Electricity supply	0.03	0.1	0.02	0.1
Construction	0.04	0.8	0.24	6.8
Wholesale and retail trade	0.20	5.7	0.60	24.1
Other transport	0.00	0.0	0.03	0.4
Civil government	0.00	0.0	0.00	0.1
Other services	0.44	1.4	0.42	2.0

10.5. Results of the Scenario Analysis

The results of the policies in the various scenarios are discussed for the production sectors, the household groups, employment and trade. In all scenarios the revenues are redistributed by means of lowering the labour tax that employers pay, i.e. a subsidy on the demand for labour by the production sectors. The following levies as described in Section 10.5 are imposed in Scenarios 1 to 5.

1. A regulatory levy on the primary use of zinc (base scenario).
2. A regulatory levy on the throughput of zinc.

3. A regulatory levy on products that contain zinc.
4. A regulatory levy on the primary use of lead.
5. A regulatory levy on the primary use of both zinc and lead.

In each of the first three scenarios, a levy is imposed on the use of zinc at different phases in the M-P chain in order to analyse the difference between the effects of an input, a throughput or a product levy. These three levies will have the same labour effects, because the revenues are equal (see Section 10.4) and redistributed in the same way. The effects of the metal levy will be different. Therefore, the total effect, i.e. the combined effect of the metal levy and the labour tax, will also differ. However, a complete analysis of these partial effects is beyond the scope of this chapter. Scenarios 4 and 5 are examined to compare the effects of a levy on zinc with a levy on lead and a levy on both metals.

The results of the scenario analyses are presented as absolute (i.e. real quantity) changes from the initial equilibrium. Tables 10.2 to 10.5 give the main results of the scenarios for selected production sectors, household groups, employment and trade. Detailed results, including the partial effects, can be found in Kandelaars and Dellink (1997). The partial effects of the metal levy and the labour tax of Scenario 1 are discussed. The partial effects of the other scenarios are analogous to those of this base scenario.

Scenario 1: A regulatory levy on the primary use of zinc

The regulatory levy on the primary use of zinc has a considerable impact on the basic metal industry: Table 10.2 shows that the real output level decreases by 10.5%. The basic metal industry accounts for over 80% of the total primary use of zinc (see Table 10.1). If this industry could transfer the tax burden to other sectors, the total effect would not be so large. Therefore, it may be concluded that the basic metal industry has only limited opportunities to transfer the tax to its buyers. This is consistent with the empirical observation that international competition in this sector is severe. Table 10.2 also shows that the main sectors affected by the policy are metal products, electricity and tobacco products.[8]

For most sectors the effects of the regulatory metal levy are small, because most products and services have only a minor metal content and the reduction in labour taxes has no major impact on the competitive position of the sectors.

The partial effect of the levy on zinc (excluding the effect of the labour tax reduction) is generally negative and small for the labour income households, and positive for the self-employed, the pensioners and the transfer recipients. It is interesting that the higher income households (for example, labour income households, pensioners and transfer recipients) are more negatively (or less positively) affected by a levy on zinc than lower income households. This implies that a levy on zinc leads to a slightly more equal income distribution. A reason for this may be that the (implicit) consumption of metal products increases more than

---

[8] The result for the tobacco industry is surprising and occurs in most scenarios. Apparently, this is positively influenced by a number of factors, each of which are of minor importance.

proportionally with income.

The effect of the labour tax reduction is positive for the 'workers', i.e. the labour income households and the self-employed, and negative for the 'non-workers', i.e. the pensioners and the transfer recipients. The difference between workers' and non-workers' households arises because workers will claim part of the labour tax reduction to increase their disposable income. The labour tax reform has a more positive effect on the high-income than on the low-income households. Therefore, this part of the scenario results in an increase in income inequality.

Table 10.2. Model results for selected production sectors; Scenarios 1 to 5.

	Real output change (in %)				
Production sector	Scenario 1	Scenario 2	Scenario 3	Scenario 4	Scenario 5
Agriculture	0.01	0.12	0.02	0.04	0.02
Tobacco products	0.43	1.16	0.49	0.42	0.42
Petroleum refineries	0.18	0.37	0.25	0.12	0.16
Basic chemical industry	-0.09	-1.62	0.46	-0.13	-0.10
Basic metal industry	-10.51	-0.07	-13.15	-7.01	-9.24
Fabricated metal products	-0.56	-3.41	-0.54	-0.95	-0.71
Electricity supply	-0.46	-0.41	-0.55	-0.37	-0.43
Construction	-0.01	-2.99	-0.01	-0.33	-0.13
Wholesale and retail trade	-0.10	-0.11	-0.07	-0.29	-0.17
Insurance	0.06	0.11	0.07	0.05	0.05
Civil government	0.12	0.40	0.13	0.15	0.13

The first column in Table 10.3 shows the total effect (i.e. the combined effect of the material levy and the labour tax reduction) of Scenario 1. It appears that the policy has a positive impact on the workers and the transfer recipients. The positive impact is only significant (around 0.25% of their net income) for the self-employed household groups: both the partial effects of the metal levy and the labour tax are positive for these groups. This scenario has a negative, but minor impact on pensioners. For most households the effects of the metal levy and the labour tax reduction are opposite, which implies that the policy effects balance each other to a large extent. Only for the self-employed are both effects positive. The total effect of this scenario shows a decrease in income inequality, i.e. the low-income groups benefit more from this scenario than the high-income groups.

Table 10.4 shows that the demand for labour decreases slightly in Scenario 1, which indicates that there is no double dividend. The labour tax reform has a positive effect on the import of primary inputs, but the zinc levy off-sets this. Imports decrease, while exports (the demand of the 'Rest of the world') are affected slightly negatively by the regulatory material levy (see Table 10.4). Thus, the effects of the regulatory material levy on the trade balance are small and inconclusive.

Using the rough measure for the effects on materials (see Section 10.2), the total use of zinc in Scenario 1 will decrease from 270 to 247 kilotonnes. This is a decrease of 8.5%. Comparing this to the overall decrease in output (-0.2%), it is clear that the policy produces a (zinc) dematerialization of the economy.

Table 10.3. Model results for selected household groups; Scenarios 1 to 5.

	Real output change (in %)				
Household group	Scenario 1	Scenario 2	Scenario 3	Scenario 4	Scenario 5
Public sector	0.14	0.57	0.15	0.19	0.16
Labour income; 1 person; 1st income quartile	0.11	-0.20	0.10	0.07	0.09
Labour income; 1 person; 2nd income quartile	0.10	-0.43	0.08	0.04	0.08
Labour income; 1 person; 3rd income quartile	0.05	-0.46	0.03	0.01	0.04
Labour income; 1 person; 4th income quartile	0.03	-0.53	0.02	-0.01	0.02
Labour income; more persons; no children; 2nd income quartile	0.09	-0.34	0.08	0.04	0.07
Labour income; more persons; with children; 2nd income quartile	0.10	-0.30	0.08	0.05	0.08
Self-employed; services	0.24	0.17	0.30	0.12	0.19
Pensioners; 1 person; 2nd income quartile	-0.03	-0.20	-0.03	-0.05	-0.04
Pensioners; more persons; 2nd income quartile	-0.01	-0.10	-0.01	-0.02	-0.02
Transfer recipients; 1 person; 2nd income quartile	0.02	-0.06	0.02	0.01	0.01
Transfer recipients; more persons; 2nd income quartile	0.01	-0.06	0.01	-0.00	0.01

Scenario 2: A regulatory levy on the throughput of zinc

A levy on the throughput of zinc negatively influences the output of most production sectors. The associated labour tax cut has a slightly positive effect, while the total effect is negative for these sectors. Table 10.2 shows that this scenario has a negative effect of more than 1% of the output of some sectors: for instance, the metal products, construction, and the wood and furniture industry. Compared with Scenario 1 the main differences are that the basic metal industry is hardly affected, and that most other production sectors are affected more severely than in the first scenario. The reason is that in this scenario the tax burden is spread more evenly over the production sectors.

Table 10.4. Selected model results for employment and trade; Scenarios 1 to 5.

	Changes in demand (in %)				
Primary inputs	Scenario 1	Scenario 2	Scenario 3	Scenario 4	Scenario 5
Medium-paid labour supply	-0.0043	-0.0085	-0.0038	-0.0043	-0.0043
Competitive imports	-0.58	-1.13	-0.54	-0.52	-0.56
Exports	-0.09	0.17	-0.10	-0.05	-0.08

Note: The competitive imports do not include the import of crude petroleum and natural gas. Exports are the demand of the 'Rest of the World'.

For the households with a labour income, the effects of the regulatory zinc levy are much more negative than in the base scenario. The reason may be that now more labour intensive sectors carry the tax burden, instead of the more capital intensive basic metal industry. For the other household groups there is only a small, but negative change. The total effect is negative not only for the pensioners, as in the base scenario, but also for the households with a labour income and most transfer recipients.

This scenario has a negative effect on the demand for medium-paid labour (see Table 10.4). Therefore, there is no double dividend. The effect of this regulatory levy is negative for imports and positive for the exports (see Table 10.4). Hence, the trade balance improves.

Comparing Scenarios 1 and 2 it is remarkable to observe that the sectoral and distributional effects differ considerably. For the production sectors, the difference is to a large extent related to the demand for inputs from the basic metal industry. In the base scenario, the basic metal sector is levied severely, and subsequently a part of the levy is transferred to the purchasers of its products. In this way, the 'down-stream' producers are levied indirectly. In Scenario 2, the down-stream producers are levied directly on the basis of their metal intensity. Consequently, the metal-intensive sectors (generally the industrial sectors) are negatively affected, while the less metal-intensive sectors (e.g. basic industries and services) can increase their output. This substantiates the conclusion that the basic metal industry is not very capable of transferring its imposed extra costs to the other sectors. In the base scenario, most metal intensive sectors (see Table 10.1) can even increase their output, while in the second scenario their output decreases.

In the base scenario the effects on households were negligible, but in Scenario 2 this is no longer the case. A reason may be that now the material levy is more spread among the metal using production sectors which results in more opportunities for transferring the costs to final consumers. Moreover, in this scenario the more labour intensive sectors are levied.

The total reduction in the use of materials in the production process is smaller than in the base scenario. A reason for this may be that the basic metal industry is much less affected, and hence its use of materials is much less reduced. Moreover, in this scenario the substitution effects that are not accounted for will presumably be larger.

Scenario 3: A regulatory product levy on zinc
The results of a product levy on the production sectors in Table 10.2 show that the basic metal industry is strongly affected, with an output reduction of 13%. This reduction is larger than in Scenario 1. It may therefore be concluded that a product levy is worse for the basic metal industry than a levy on the primary use of zinc which had a negative effect of 10.5%. The reason is that in this scenario only the input from the basic metal industry is levied. Although this levy is imposed on the buyers of the products of the basic metal industry, they may switch their demand to other products, which affects the basic metal industry negatively. As well as the basic metal industry, the metal products and the electricity sectors are negatively affected. A reason may be that their possibilities of using alternative materials/products is limited. The tobacco industry and the basic chemical industry, however, are positively influenced sectors. The basic chemical industry, which has a high zinc intensity, was negatively affected by the policy of Scenario 1, but in this scenario the effect is positive. This may be due to the fact that the basic chemical industry has a relatively high inflow of materials (see Table 10.1) which are not levied here. Another reason may be that the intermediate production sectors are substituting zinc for plastics, so that the demand for products of the basic chemical industry increases due to the levy on zinc. Apart from the basic metal industry and the basic chemical industry the results are comparable with those of the base scenario. In both scenarios, the results are to a large extent dominated by the negative impact on the basic metal industry.

For most household groups this scenario and the base scenario have similar results (see Table 10.3). This similarity is interesting because a product levy affects the household groups directly, in contrast to the base scenario. This demonstrates a 'normal' mechanism of AGE models.

Table 10.4 shows the effects of this scenario on the trade balance and they are again inconclusive. Imports of crude petroleum increase, but the other imports decrease. These effects have the same sign as in the base scenario, but they are larger. The employment effects are negative, but insignificant, just as in the base scenario.

The large decrease in the output of the basic metal industry is reflected in a relatively large decrease in the total use of materials. However, the crude assumption that there is no inter-sectoral substitution is of importance here. After all, if the use of zinc itself is taxed, the producers may substitute other metals for zinc, inducing a reduction of zinc use at a constant input from the basic metal industry. If the input from the basic metal industry itself is taxed, regardless of the zinc content of that input, there will be no incentive for the producers to switch to other metals.

Scenario 4: A regulatory levy on the primary use of lead
This scenario differs from the base scenario (Scenario 1) in that the levy is imposed

on lead and not on zinc. The results for the production sectors show that the output of the basic metal industry will be reduced by 7%, which is less than in Scenario 1 (see Table 10.2), because the share of the basic metal industry in the total use is smaller for lead than for zinc (see Table 10.1). The impact of the levy on lead on various other sectors is around 0.3 to 0.4%. The spread of the lead intensities over the production sectors is larger than the zinc intensities (see Table 10.1), which implies that the tax burden is more evenly distributed over the various production sectors that are using lead. This more even distribution of the tax burden implies that the policy affects more production sectors negatively. Therefore, for a large number of production sectors, the effects are more negative or less positive than in the base scenario.

The impact of a levy on lead on household groups is roughly the same as in the base scenario (see Table 10.3), with the exception that the income improvement of labour income households is smaller than in the base scenario. A reason for this may be that a lead levy is more dispersed and more labour intensive production sectors are subjected to the levy. Here, the positive impact of the labour tax reform is completely off-set by the negative impact of the levy on lead.

The effects on imports are negative (see Table 10.4). For competitive imports, the decrease in imports is slightly larger than in the base scenario (see footnote 2). The decrease of non-competitive imports is slightly smaller than in the base scenario. The employment effects are similar to those in the base scenario. Again, there is no double dividend.

The total use of lead will decrease by 3.5%. This decrease is smaller than the decrease of zinc in the base scenario. A reason for this is that the total tax burden is spread more evenly over the production sectors. This may be clarified by looking at the demand and the input substitution effects of a policy. The demand substitution effect is the effect caused by the buyer of products from a production sector that is levied. The input substitution effect occurs when a production sector that is levied seeks opportunities for substituting the input that is levied. In this scenario, both these substitution effects may be smaller than when the levy is imposed more heavily on one sector (as in Scenario 1), because the total tax burden is spread more evenly over the various sectors. As indicated before, substitution possibilities were rather small for the basic metal industry.

Scenario 5: A regulatory levy on the primary use of both zinc and lead

This scenario is studied in order to obtain an idea of the combined effect of levies on both zinc and lead. This combined levy can give an insight into the economic effects of a 'more complete' metal policy. Policy makers who are thinking about implementing a material policy may want to impose it on several materials simultaneously. They may, for example, choose to impose a higher levy on more harmful materials. To analyse such a policy in this scenario a combined policy on zinc and lead is implemented. This scenario may be compared with Scenarios 1 and 4 where a policy is imposed on only one metal.

Table 10.2 shows that the effects of this scenario are between those of Scenarios 1 and 4. If in Scenario 1 (4) zinc (lead) was substituted by lead (zinc), it would be expected that in a scenario where both metals are levied the costs of substitution and the output effects would be higher, as some of these substitution possibilities are no longer available. This effect does not occur, and therefore it may be concluded that there is not much substitution between zinc and lead. Also for the household groups, employment, trade and total use of materials, the effects of this scenario are somewhere between those of Scenarios 1 and 4 (see Tables 10.3 and 10.4).

10.6. Conclusions

This chapter has presented a first 'empirical step' in filling the gap between physical and economic models, by combining a material flow model and an applied general equilibrium model. In this way, the effects of material policies may be analysed from both environmental and economic perspectives. The model is applied in order to study the impact of policies to reduce the use of zinc and lead in the Netherlands.

However, the current formulation of the integrated model is far from perfect. There are no endogenous feedback mechanisms from the economic model to the material flow model, and the substitution between materials within one production sector cannot be accounted for, because the structural characteristics are exogenous and fixed. However, substitution between materials from different production sectors is taken into account. Nevertheless, even from the current analysis some qualitative conclusions can be drawn.

First, a regulatory levy on the use of materials will have only minor effects on most production sectors and households. The macroeconomic effects are likely to be small. The main economic effects are the output losses of those sectors where most materials enter the economy. On the other hand, the environmental benefits are clear: total use of materials will decrease significantly.

Specifying an alternative scenario, where the first stage production sectors are exempt from the levy, will have a more dispersed effect on the economy. More production sectors are affected by the regulatory levy, but none to the extent of the first stage production sectors in the base scenario. However, the environmental gains will also be smaller.

Finally, the economic effects of a product levy (irrespective of the use of materials) will be similar to those of a material levy. Unfortunately, the differences in the effects on the use of materials between a material and a product levy cannot be assessed accurately in this version of the model.

From this illustration of material policies in an economic model it may be concluded that the combination of an AGE model and a material flow model is an appropriate tool for analysing the sectoral, distributional, and environmental effects of material policies. Consistent analysis of the interactions of these effects is important to inform environmental-economic policy makers.

More research is needed to more accurately assess the material and economic impacts of policies in an integrated physical and economic model. Two further steps towards this integration could be the inclusion of a direct feedback from the AGE model to the material flow model and the modelling of dynamic effects. In this chapter the feedback between the AGE model and the material flow model is exogenous. Moreover, the material intensities are assumed not to change when a levy is imposed. These weakness may be overcome by integrating the AGE model and the material flow model, and by allowing for an endogenous feedback between these two models. This endogenous feedback may be implemented by means of a calibration of the AGE model, using a new economic sector (perhaps called 'environment') that supplies the materials to the production sectors. In this way, the material flow model will essentially become a part of the AGE model. This way of modelling also provides many new possibilities, as other environmental themes, like emissions, can be included in the same manner as materials.

With a dynamic model the impacts over time and the role of technological development may be analysed. Technological development may be a major source for reducing the use of materials. For instance, material policies may produce endogenous changes in technology, as in recycling, more material-efficient production techniques and new substitution possibilities.

CHAPTER 11

SUMMARY, CONCLUSIONS AND PROSPECT

11.1. Summary

Research on material flows in an economy-environment system has hitherto mainly been performed in environmental science and has focused on (i) describing the physical flows required for producing a particular product, or (ii) describing physical flows in a certain period and region. Little attention has been devoted to physical flows in environmental economics. The present study attempts to fill this gap between environmental science and environmental economics. The goal of this study is to examine the physical and economic mechanisms related to flows of materials and products, as well as the policies and strategies designed to reduce resource scarcity and environmental pollution related to these flows. Thus, this book aims to contribute to integrated model-based analyses of resource and pollution problems for policy making. Although this theme of policy analysis and material flows is not entirely new, in the present study it has been approached with novel insights, concentrating on interlinked material and product flows, associated environmental issues, and substitution and recycling mechanisms in production and consumption processes. As a general framework the concept of a 'material-product (M-P) chain' has been adopted. This is defined as a system of linked flows of materials and products supporting the provision of a certain service. This means that it includes flows of one or more materials and flows of one or more products. The term 'chain' refers to the sequence of economic activities required to provide that service. This sequence consists of extraction, material production, production of products, recycling, reuse and waste treatment. Chain management can be linked to this as an overall policy strategy that explicitly considers sequential linkages between various activities in terms of both economic and physical mechanisms.

Using the concept of an M-P chain various analyses can be performed. In this study 'economic analysis of M-P chains' is defined as the study of allocation and substitution processes in an M-P chain. Economic analysis of M-P chains may be based on models of static or dynamic optimization, market equilibrium and policy analysis, and may include substitution at different levels (material, production, product).

This study has examined theoretical, modelling and applied issues in M-P chain analysis. The two main objectives were: (1) the formulation of various types of economic models of M-P chains in order to analyse the combined economic and physical impact of environmental policies; and, (2) the application of these models to various empirical M-P chains. The study was divided into three parts. Part I, comprising Chapters 1 to 3, covered conceptual issues of material flows from environmental and economic perspectives. Part II, including Chapters 4 and 5, reviewed physical flow models and economic models that may be used to study interactions between material and product flows. Part III, consisting of Chapters 6 to 10, discussed applications of M-P chain analysis.

Basic concepts of physical flows in environmental and economic systems were examined in Chapter 2. The notions of ecosystems, material cycles, metabolism, evolution, the laws of thermodynamics and the material balance (MB) principle were discussed. Some of these concepts apply to both environmental and economic systems (e.g. material cycles, metabolism, laws of thermodynamics and MB conditions) while others can be used as analogies to economic systems (e.g. ecosystem, evolution). Chapter 3 discussed two distinct approaches to environmental policy problems, both of which were applied later in this study. The first is a welfare (monodisciplinary) optimizing approach focusing on externalities. The other is a multidimensional (multidisciplinary) approach in which the analysis and evaluation may be based on various dimensions and criteria. Next, an overview was given of strategies and policies to reduce the environmental impact of M-P chains. The chapter finished with a survey of practical policies related to materials. Part I, the conceptual part, provided the basic ingredients for the theoretical models and applications in Parts II and III of the study.

The model types described in Chapters 4 and 5 formed Part II of the study. Chapter 4 discussed the most commonly used model types for physical flow and environmental analysis: 'material flow analysis' (MFA); 'life-cycle assessment' (LCA); and, physical input-output (I-O) analysis. These were described, analysed and classified in terms of various modelling aspects. The concept of M-P chains and the economic analysis of M-P chains was described. The four model types described have in common that their basis is the description of inputs and outputs of economic or environmental stages. MFA and physical I-O analysis describe material flows through the economy or the environment. A difference between M-P chain analysis and both MFA and physical (I-O) analysis is that the latter two do not include products. LCA studies an M-P chain because LCA examines an economic structure of material and product flows. LCA is focused on the environmental impact of products. LCA can be seen as an 'environmental' analysis of an M-P chain, and M-P chain analysis, as defined here, as an economic analysis of an M-P chain. From a physical viewpoint M-P chain analysis combines elements of MFA, physical I-O analysis and LCA.

This survey shows that MFA, LCA and physical I-O analysis are based on an input-output framework, which does not include monetary, allocative or behaviourial aspects. For an analysis of policies aiming at the reduction of environmental problems related to physical flows, it is crucial to allow for substitution, recycling and reuse, and to incorporate dynamic aspects, such as technological change. In an input-output framework these strategies are not incorporated. Therefore, a more general M-P chain analysis approach combines physical with economic aspects, and includes physical aspects less rigidly than an input-output framework.

Chapter 5 gave an overview of economic model types that may be used for the analysis of physical flows. Five types of economic models were described in relation to material flows: economic models of natural resources; pollution models;

environmental input-output models; macroeconomic models; and, models of technological change and evolution. Economic models of natural resources and pollution models are based on neoclassical economic theory. The first three model types were selected because they are often used and applied in environmental-economic modelling. Economic models of natural resources focus on the optimal allocation of resources (and possibly pollution), usually over time. With pollution models based on externalities the optimal regulation of (extraction and) pollution may be analysed. Environmental input-output models describe material flows between economic sectors and environmental sectors. Macroeconomic models, based on relations between aggregate variables, often at a national scale, are not often used for environmental issues. However, they may provide general scenarios for studies using other types of models formulated at a more disaggregated level and perhaps smaller spatial scale. Models of technological change and evolution are even less common in economic models of physical flows, but they may be useful in studying dynamic aspects, especially long-run and structural changes of material flows through an economy, such as delays between the invention and the adoption of a technology.

The main aspects of material flows that need to be integrated in economic models are: substitution, recycling and MB conditions. Substitution between materials is possible in economic models of natural resources, pollution models and models of technological change. In I-O and macroeconomic models substitution is given little attention because of fixed technological coefficients in I-O models and the high level of aggregation in macroeconomic models. Recycling is possible in all five types of models. MB conditions may be included in all the described models, except for macroeconomic models, again for reasons of aggregation. In all types of models different material flows are rarely integrated.

From this short overview it may be concluded that each of the modelling types has a particular focus and can lead to answers to particular questions. The integration of material flows in each model type differs too. The choice for a specific type of model depends on the questions that are asked.

M-P chain analysis is useful for examining the effects of various instruments or policy packages on the economy, physical flows and the environment and their interactions. For this purpose a physical flow model needs to be combined with an economic analysis. The physical flow models of Chapter 4 together with the economic models of Chapter 5 constitute the basis of the applications of M-P chain analysis in Part III of the study. Each of the applied models in Chapters 6 to 10 combines elements of these physical flow and economic models.

Chapter 6 presented a static optimization model of an M-P chain in which an 'environmental manager' optimizes the costs under a set of physical and technological restrictions. The model optimizes the costs for the demand of a service which includes, besides recycling and reuse, demand and production functions. The goal of this chapter was to explore how policies or strategies that are applied to different stages of an M-P chain may differ in their impact on a number of physical and economic indicators. The model includes MB conditions for each activity and process. The MB conditions are important to determine the amount and the type of waste material that is generated after a product is disposed of.

The environmental manager is supposed to minimize the total costs of satisfying the demand in an M-P chain. This model is a combination of a material flow analysis (MFA) (or physical I-O analysis) and elements of a pollution model (environmental manager, waste/pollution). The 'total costs' are defined as the costs of new products, reused products and waste dumping. It is possible to include external costs as well. In this model, the production process has as inputs two types of materials that may be new or recycled, and as outputs new products and waste material. The waste material from the production process may be recycled together with the waste material resulting from the disposal of the product. The choices that need to be made concern: which and how much new and recycled materials of each material in the production process; how many new and reused products for meeting the demand for the service; and, which percentages of the materials will be recycled. MB conditions are included on the level of the product, the production process and the system. The MB conditions are important to keep track of the material content of products and the waste material that is generated after a product is disposed of. The model may include endogenous prices and different production technologies. With this type of model the optimal recycling and reuse rate, and the optimal input mix of materials in the production function may be assessed. Furthermore, the impact of policy instruments (including physical restrictions, such as the minimal rate of recycling) on material dumping, recycling, reuse of products and the costs of meeting the demand can be analysed.

The model was applied to rain gutters, where a distinction was made between zinc and pvc gutters. The basic optimization model is linear. The production functions are linear, with one input resulting in a fixed amount of waste material for each type of gutter. All waste material is recycled as long as the recycled material is cheaper than the new material, otherwise no recycling takes place. These extremes set the percentages of materials that are recycled. In a linear model the optimal allocation of the demand over the two types of rain gutters is either only zinc or only pvc gutters. Alternative policies, such as a product charge, recycling standards and subsidies, may change the distribution of demand resulting in a change in the extraction of materials and the total costs of the M-P chain. In a non-linear model the demand is met by a combination of zinc and pvc gutters. In the scenario in which the price of waste treatment depends on the number of zinc gutters (which makes the model non-linear), the demand is met by both zinc and pvc gutters. In this scenario, the levy on waste treatment (i.e. a levy on the dumping of galvanized steel) increases the net costs and a certain proportion of the zinc gutters will be substituted by pvc gutters. The results show that the current situation is not optimal, in the sense that the net costs of the demand are not at their minimal level.

The application showed that for an analysis of policies related to materials or products, the demand for products and materials needs to be considered simultaneously. A policy imposed on a product or material may affect the M-P chain at different, but connected levels: for instance, the material, product and demand level. The model minimizes the total costs of the system for different (policy) scenarios, and it may be used to analyse the differences in total costs, extraction and waste disposal of each (policy) option.

Chapter 7 performed a general equilibrium analysis of an M-P chain. In that model the demand and supply of all economic agents are matched on markets, resulting in a price for all products, materials and capital. In a partial equilibrium model some markets would have fixed prices. The general equilibrium model is not an economy-wide model, but a model for a specific M-P chain. The model includes a new approach, based on externality theory, that integrates MFA and environmental policy analysis. In other words, a welfare-economic perspective is integrated with a physical perspective. The model combines extraction, production, recycling, consumption, waste treatment activities and MB conditions, in a general equilibrium framework. Although some studies have mixed some of these elements, the total combination, as pursued in this chapter, is new. A social welfare optimum is defined in which externalities are included. In the market equilibrium these externalities are optimized by imposing adequate taxes.

The results show that the externalities caused by extraction and generation of harmful waste can be optimized by imposing a tax on new (raw) materials. In a second-best world these externalities may be optimized by imposing taxes on harmful waste and on the use of recycled materials. The optimal taxes on the generation of harmful waste and on new materials depend partly on the same term that includes prices and marginal products of production and waste treatment functions. This implies that a change in some variables causes a shift of taxation from the start (extraction activity) to the end (waste treatment activity) of the chain. This linkage implies that the whole M-P chain needs to be considered when analysing optimal policy packages. Therefore, to derive the optimal rules for taxes and subsidies it is necessary to consider flows and processes related to raw materials, recycled materials, main products, garbage from consumption, and material and recycling waste.

This model includes MB conditions for every economic activity in the M-P chain. In a production function with various inputs, MB conditions are important to keep track of the material content of products. After a product is disposed of the products need to be transformed back into the one or more materials of which they were originally made. If more types of materials are used, it is relevant to distinguish between various types of waste material. When a production function has several output, for example, products and production waste, MB conditions are needed to determine the material content of the products and the amount of production waste. Thus, a multi-input or multi-output function (e.g. a production, recycling or waste treatment function) requires that for each type of material (or product) the input equals the output.

In this equilibrium model the relevance of including the M-P chain, from extraction to waste treatment and the linkages between all stages, is apparent from the variables in the optimal tax rules. This general equilibrium model, which includes physical properties of production and consumption, provides insights about the connection between externalities generated by extraction and waste treatment.

Chapter 8 presented a dynamic descriptive model to study the impacts of economic and policy developments aimed at reducing the use of materials. This model is a combination of MFA (physical I-O analysis) and elements of an economic model of natural resources (e.g. allocation over time). In this dynamic analysis temporal issues, such as accumulation and delays, are taken into account.

Accumulation of products and materials may occur in the economy: for example, for durable products there is a time-lag between the purchase and the disposal of the product. Delays may occur between the implementation and the effect of a policy. The model traces the effects of changes in material flows over time: for example, through substitution and recycling. Using the model, policy scenarios were simulated to assess their influence over time on an M-P chain. The model was applied to rain gutters.

Rain gutters are durable products with a lifetime of several years. To distinguish between products that are produced in different years, a vintage approach is adopted to describe rain gutters. The model describes the demand for rain gutters which depends on the number of new houses to be built and the number of houses to be renovated. This demand is allocated to zinc or pvc rain gutters. The model analyses the demand for zinc and pvc gutters and the material flows associated with this demand, under various (policy) scenarios. Simulation results show that a shift in consumer preference from zinc to pvc gutters leads to a reduction in the use and disposal of zinc, but to an increase in the disposal of pvc. However, over time the demand for gutters increases because the lifetime of pvc gutters is shorter than that of zinc gutters. The product charge scenario imposes a charge on zinc gutters. Subsidies on recycled zinc and pvc do not affect the demand or the allocation of gutters, but have a positive effect on the extraction of zinc ore and the disposal of zinc and pvc waste into the environment.

This dynamic descriptive model for gutters reflects the importance of products and processes for the analysis of flows of materials, tracks the impact of economic and government policy variables on M-P chains, and includes dynamic aspects such as accumulation and delays.

In Chapter 9 a dynamic analysis of M-P chains was performed in which a life-cycle assessment (LCA) was combined with an economic analysis. The dynamic descriptive model includes the environmental impacts of the M-P chain. Also accumulation and delayed effects are considered. The dynamic model was applied to window frames. The demand for window frames is allocated to four types of window frames. The distribution over the various types depends on whether the window frames are for newly built houses or for renovated houses. Window frames are durable products with a long lifetime, and therefore a vintage approach is also adopted here. Three economic agents are modelled, namely producers (of window frames), consumers (e.g. a construction firm) and a regulator (e.g. a government).

In the model two policy packages are imposed aiming to reduce (1) the depletion of raw materials, and (2) water pollution. The policy package for reducing the depletion of raw materials consists of levies that are imposed on hardwood window frames and on the dumping of hardwood, and information provided to consumers to persuade them to choose pinewood window frames. To reduce water pollution, the policy package consists of levies on the use of new aluminium and pvc, on the dumping of aluminium and pvc, and on the use of aluminium window frames. As in (1) above, information in favour of pinewood window frames is provided to consumers.

The results show that imposing a mix of policies on depletion may reduce the raw materials depletion by more than 50% in 20 years compared with the scenario without policies. Also other environmental aspects are positively affected. The policy package on water pollution has a delayed effect, because water pollution mainly occurs at the waste treatment stage of the M-P chain. After 40 years, water pollution is reduced by 30% compared with the base scenario. The package also reduces other environmental impacts substantially compared with a scenario without policies. Both policy packages have a positive effect on the average cost per window frame paid by the consumers. The revenues for the government from the levies are positive but decreasing over time, because consumers buy less of the product on which a levy is imposed. A multi-criteria analysis was performed to see the effects of single instruments on the environmental indicators. This analysis shows that with equal weights for each indicator, a levy on aluminium window frames is the best single instrument. This type of model can be used to analyse the effects of a specific policy (package), aimed at reducing a certain environmental indicator, on other environmental and economic indicators. These total effects need to be taken into account simultaneously.

Chapter 10 presented a first 'empirical step' in filling the gap between physical and economic models, by combining a material flow model and an applied general equilibrium (AGE) model that are both calibrated for the Netherlands. The goal of this model was to empirically assess the economy-wide and environmental effects of environmental policies focusing on the environment. The AGE model used for the model is the 'Taxinc-model' which is a static AGE model with 61 production sectors and 44 household groups. It allows the analysis of the sectoral and distributional consequences of policies. The material flow model 'Flux' is an input-output model in physical units that describes the material flows for each sector of the economy. Thus, Flux consists of a database on stocks and flows of materials.

To combine the Taxinc-model and Flux, the economic sectors of Flux are connected to those of the Taxinc-model. A material levy on a sector in the Taxinc-model depends on the use of materials (in kilograms, derived from Flux) and on their input or output (in monetary units, derived from the Taxinc-model). Thus, the levy combines the physical and monetary units. This approach facilitates the analysis of the effects of material and product policies on various production sectors and household groups, and on trade and employment. The changes in the use of materials are determined on the basis of these effects. In an AGE model, substitution between different production sectors is considered, but, for studying material flows, substitution within a production sector may also be important.

Material policies are imposed to reduce the use of specific materials. Such policies generate revenues, which in this model are redistributed to the tax payers by lowering the labour tax. A material levy and a labour tax reduction shift the tax burden from labour to the use of materials. This shift is attractive because it taxes more what we want less of (use of materials), and less what we want more of (labour).

The policies are formulated for the metals zinc and lead. The results show that the effects of a regulatory material levy are small for household groups and for most production sectors. The impact of this levy on the basic metal industry and some other large metal-using production sectors may be significant. In theory, the combined effect of the material policy and the labour tax reduction may have a positive effect on the environment (i.e. less material use) *and* on employment, which is called a 'double dividend'. Here, however, no double dividend was found for any policy scenario.

From this illustration of material policies in an economic model it may be concluded that the combination of an AGE model and a material flow model can produce an appropriate tool for analysing environmental, sectoral and distributional effects of material policies. Consistent analysis of the interactions of these effects may be important for environmental-economic policy making.

11.2. General Conclusions on M-P Chain Analysis

An economic analysis of an M-P chain, or M-P chain analysis, allows for an integrated analysis of resource and pollution problems from a physical and an economic perspective. Traditionally, environmental economics has mainly focused on a partial analysis of environmental problems, resulting in a neglect for the interdependence of environmental problems caused by different economic stages. Often environmental economics focuses on external effects, without considering the material or physical dimension of problems. Economic processes are linked to - and even regarded as embedded in - physical processes and therefore a change in an economic process affects the physical process and vice versa. To include this physical dimension, material flow models may be combined with economic models. This allows the study of policy packages in which physical and economic aspects are considered simultaneously. This results in analyses that are economically consistent and physically feasible.

An analysis of material flows in an economy requires the study of issues such as the recycling of materials and the reuse of products. The opportunities for reducing the environmental impact of M-P chains are considerably limited by not taking these aspects into account. The assessment of the environmental impact of a product may be quite different when recycling and reuse options are not considered. Furthermore, substitution on different levels needs to be taken into account: substitution between materials, between materials and other inputs, and between products. Without considering these issues the effects of a policy may be incompletely examined. For example, a reduction in the use and the environmental effect of one product may increase the use and environmental effects of another product that provides the same service. The environmental impact of a product may be considered, incorrectly, to be higher or lower when substitution options are not taken into account.

MB conditions need to be included for every economic activity to ensure that the economic model does not generate policy options that are physically incorrect. It is also important to keep track of the materials that are contained in a product with a view to the ultimate waste treatment of products that are disposed of. When a production process (or function) has multiple inputs or multiple outputs, it may be a problem to assess the amount of (different types of) material contained in the product. This consideration is also relevant for the transformation of products into materials, because the amount of materials that may be recycled and the quality of the materials may depend on the amount of capital or labour that is used in the transformation process.

A dynamic analysis of M-P chains makes it possible to study delayed effects, accumulation of materials or products, technological changes and other development paths. Especially when durable products are considered, a dynamic analysis is more appropriate. The effect of a policy may be delayed: for example, a policy imposed on the material used in production in order to change the amount of waste generated by disposed products, may have a delayed effect when the product has a lifetime of several years (i.e. a durable product).

Chain management is based on a mix of instruments that need to be attuned, given the environmental or external effects of each activity in a chain. In M-P chain analysis the linkages of particular activities between resource extraction and waste treatment are considered, allowing also indirect effects of policies to be considered. For instance, a reduction in the use of one material to reach a certain level in terms of environmental indicator X may require less use of a particular product, but then due to a resulting increase of the use of another product, which provides the same service, the use of another material may increase and environmental indicator Y may be negatively affected. These are difficult trade-offs, but they can only be made explicit after the different physical and environmental dimensions are linked to each other via economic mechanisms. In theoretical or analytical economic models of M-P chains these linkages may be included by, for example, extending economic models. For empirical models this is more complicated, because information is required on the behaviour and choices of economic agents with regard to products, materials, recycling and waste treatment. Furthermore, models are required for recycling activities and waste treatment.

In an M-P chain analysis, various economic activities are connected because the output of one activity can be the input of another: for example, products are the output of production and the input of consumption. Recycling and reuse activities form 'loops': for instance, materials go from consumption to production via the recycling activity. For M-P chain analysis these connections are important, because the amount of materials extracted and waste generated are reduced when more recycling takes place.

11.3. A Comparative Evaluation of M-P Chain Models

The goal of this study was to formulate and apply various economic models to analyse M-P chains. This section presents a comparative evaluation of the various models that were applied in Part III of this study. This is relevant for deciding which model can best be used for which purpose. Table 11.1 summarizes the main differences between the models.

With an M-P chain analysis environmental and policy issues may be studied. For selecting a certain type of model it is important to clearly define the issues that need to be studied and match these with the characteristics of the different types of models including their purposes (Kandelaars and Van den Bergh, 1998).

An optimization model may be chosen when the focus is on minimizing costs or environmental impact. Chapter 6 presented an optimization model in which the costs are minimized. With this cost-effectiveness model the effects of various policies for one service are analysed, but without considering their effects on other parts of the economy. The optimization may be in terms of least costs to meet the demand under a set of physical or environmental constraints, or in terms of minimizing the environmental impact under a set of economic and physical constraints. Physical constraints are, for example, MB conditions or production functions. The purpose of this type of model is to analyse the interactions between economic, physical or technological aspects of an M-P chain and the effect of (environmental) policies on material and product flows.

If a model is needed to calculate the optimal tax rules or the effects of certain taxes on various markets, an equilibrium model may be used. Such a model describes an equilibrium set of prices and quantities on each market. With this type of model a market equilibrium with an implemented policy may be compared with a social optimum, thus including externalities. Chapter 7 presented a static general equilibrium (GE) model that includes two types of externalities, directly related to extraction and pollution, respectively. The model determines the optimal tax rules in a market equilibrium in which the externalities are 'internalized'. This analytical model describes various economic activities for one M-P chain.

Table 11.1. Comparison of economic models for M-P chains.

Chapter / *Aspect*	6	7	8	9	10
Focus	Optimization of the costs of demand	Optimal taxes in a market equilibrium	Delays and accumulation	Delays and accumulation; environmental impacts	Effects on production sectors and households
Approach	Multi-dimensional	Welfare-economic	Multi-dimensional	Multi-dimensional	Welfare-economic
Impact of physical flows	Physical	Physical and externalities	Physical	Physical and environmental	Physical
Temporal aspect	Static	Static	Dynamic	Dynamic	Static

An equilibrium model may also be used to analyse the direct and indirect effects of policies on various production sectors and household groups in an economy. Chapter 10 presented an applied general equilibrium (AGE) model which is calibrated for the Netherlands. The AGE model is combined with a material flow model for the Netherlands by connecting the production sectors of both models. The physical data on material flows between economic sectors are connected to data in monetary units in the AGE model. With such a model one can determine the effects of material policies on economic sectors. This model is a tool to analyse the sectoral, distributional and environmental effects of material policies. An AGE model is appropriate when it is expected that a policy has economy-wide effects and affects various sectors.

The GE models of Chapters 7 and 10 have in common that each market clears and that prices are determined within the model. Both models combine an economic model with physical aspects. Both models are static and do not include dynamic aspects. The analytical GE model of Chapter 7 includes MB conditions in each stage, thus also in the phase of recycling and waste treatment activity. Chapter 10, the AGE model, includes many different material flows of which only one (or several) are connected to the physical material flow model. The MB conditions only hold for the physical flows that are studied. The model in Chapter 10 includes distributional effects between different household groups, while in the Chapter 7 model all consumers are identical.

The AGE model is empirical and shows the economy-wide and distributional effects, while the analytical GE model focuses on the effects of (optimal) taxes within one M-P chain. The models may be used as complements: the optimal taxes may be derived with the analytical model, and with the AGE model these taxes may be analysed in an empirical model.

Dynamic models are used to examine the past or the present. On the basis of such a description future developments or changes may be analysed possibly with the effects of various policies. Dynamic models are particularly suitable for analysing M-P chains of durable products, such as the rain gutters and window frames that were considered in Chapters 8 and 9, respectively. Policies or technological change may have delayed effects, for example, the time-lag between the invention and the adoption of a technology. Changes over time, for example, due to changes in preferences are considered in dynamic models. For studying material and product flows, accumulation needs to be considered. Accumulation may occur at different levels. At a product level accumulation takes places when there is a time period between the purchase and the disposal of a product, and at the level of an M-P chain (or an economy) when there is a time period between the extraction and disposal of materials from or into the environment.

With these dynamic descriptive models one can project the possible effects over time of technological or economic changes or policies. Chapter 8 provided a descriptive model in which economic and physical aspects of an M-P chain of a durable product were analysed for a number of (policy) scenarios. In Chapter 9 the dynamic analysis included environmental impacts and the evaluation of policies particularly directed at reducing environmental effects.

The economic models of M-P chains which were formulated and applied in this study can be used to analyse the economic, physical and environmental impacts of environmental policies. With these alternative models the analysis of the interactions between material and product flows and economic models can be improved. As described in Section 11.2, their improvements compared to existing models for analysing material and product flows are: the integration of resource and pollution problems; the incorporation of substitution at various levels, the recycling of materials and reuse of products; the analysis of the connections between various economic stages; MB conditions are included at every stage and level in the model; and, the combined analysis of the economic, physical and environmental effects of various policies, technological changes or other developments.

11.4. Prospect

Several research questions may be addressed that build on the results of this study, in order to improve the models in an analytical sense or to make these models more appropriate as tools in environmental-policy analysis.

Further research on the behaviour of economic agents in their choice for materials, products, recycling, reuse and waste treatment is required to understand the effects of policies aimed at changing this behaviour. The connection between various stages in the M-P chain, e.g. production, consumption and recycling, needs to be studied to analyse the trade-offs between different stages of the M-P chain.

Consumers and producers can choose between different (new and reused) products, and various (new and recycled) materials. These choices may require the inclusion of the notion of imperfectly substitutable products and materials in the models. Substitution between heterogenous products and materials needs to be further analysed.

Production functions which represent the transformation of materials into products with the use of capital and labour do not generally take MB conditions into account. In the models of this study these MB conditions are considered explicitly. In most traditional economic models the material contents of products are not measured. However, in an M-P chain the material contents of products are important: for example, to derive the amount of waste material generated after the product is disposed of. When a production function has multiple inputs or multiple outputs, the amount of products does not determine the amount of materials used. Therefore, with multiple inputs or outputs, it is required to keep track of the material input and output in the production function. When a product contains several materials, the input of these materials into the production function and the possible production waste has to be assessed in order to transform the product into the different types of waste material.

Further research on the dynamic and evolutionary aspects of M-P chain modelling may usefully focus on the inclusion of technological changes, new (imperfect substitutable) materials and products. Dynamic effects related to the accumulation of materials and products in the economy, for instance, due to durable products, and delays in the effects of policies, technological change or other changes, need to be studied in detail to avoid unforeseen difficulties.

With the modelling of M-P chains, the effects of policies may be analysed for various economic activities in order to study the trade-offs between an activity, an environmental problem or a geographical area and another activity, problem or area. In this study, the latter spatial trade-offs are not included. Spatial aspects that particularly require analysis are the geographical scale of imposing policies, and trade-offs between spatial levels: for instance, the shift from generating waste in the Netherlands to other countries.

More empirical research is required regarding the interactions between: economic and physical aspects of M-P chains; the behaviour of economic agents regarding the choice of materials, products, recycling, reuse waste treatment; the relationship between economic activities and environmental problems; and, the trade-offs between the economy and the environment. Empirical research may help to fine tune the models to make them more appropriate as decision-support tools for environmental policy analysis.

References

Alcamo, J., R. Shaw and L. Hordijk (eds.) (1990). *The RAINS Model of Acidification, Science and Strategies in Europe*, IIASA or Kluwer Academic Publishers, Dordrecht.

Allen, P.M. (1996a). Man and the environment: the complex systems approach, Paper presented at the Conference of the Applied Econometrics Association, Lisbon, April 11-12.

Allen, P.M. (1996b). Evolutionary complex systems and sustainable development. In: M. Hofkes and J.C.J.M. van den Bergh (1996). *Economic Modelling of Sustainable Development - Between Theory and Practice*, USF International Symposium, 20 December, Vrije Universiteit, Amsterdam.

Amir, S. (1994). The role of thermodynamics in the study of economic and ecological systems, *Ecological Economics*, Vol. 10:125-142.

Anderberg, S., B. Bergbäck, and U. Lohm (1989). Flow and distribution of chromium in the Swedish environment: a new approach to studying environmental pollution, *Ambio*, Vol. 18:216-220.

Annema, J.A., H. Booij, L. Paardekoper, L. van Oers, E. van der Voet and P. Mulder (1995). Stofstroomanalyse van zes zware metalen - gevolgen van autonome ontwikkelingen en maatregelen (Material flow analysis of six heavy metals - consequences of autonomous developments and policies), RIVM report no. 601014010, Bilthoven, the Netherlands (in Dutch).

Ayres, R.U. and A.V. Kneese (1969). Production, consumption and externalities. *American Economic Review*, Vol. 59:282-297.

Ayres, R.U. (1972). A materials-process-product model. In: A.V. Kneese and B.T. Bower (eds.). *Environmenal Quality Analysis - Theory and Method in the Social Sciences*, Resources for the Future, Washington DC.

Ayres, R.U. (1978). *Resources, environment, and economics: applications of the materials/energy balance principle*, John Wiley and Sons, New York.

Ayres, R.U., V. Norberg-Bohm, J. Prince, W.M. Stigliani, and J. Yanowitz (1989). Industrial metabolism, the environment, and application of materials-balance principles for selected chemicals, IIASA Laxenburg Austria.

Ayres, R.U. (1989). Industrial metabolism. In: J.H. Ausubel and H.E. Sladovich. *Technology and environment*, Washington, National Academic Press.

Ayres, R.U. and V. Norberg-Bohm (1992a). Industrial metabolism of sulphur, Working Paper 92/58/EP, INSEAD Fontainebleau.

Ayres, R.U. and V. Norberg-Bohm (1992b). Industrial metabolism of nitrogen, Working Paper 92/57/EP, INSEAD Fontainebleau.

Ayres, R.U., L.W. Ayres and J.A. Tarr (1994). A historical reconstruction of carbon monoxide and methane emissions in the United States. In: R.U. Ayres and U.E. Simonis (eds.). *Industrial Metabolism, Restructuring for Sustainable Development*, United Nations University Press, Tokyo, Japan.

Ayres, R.U. (1994). Industrial metabolism: theory and policy. In: Ayres, R.U. and U.E. Simonis (eds.). *Industrial Metabolism, Restructuring for Sustainable Development*, United Nations University Press, Tokyo, Japan.

Ayres, R.U. and L.W. Ayres (1994). Consumptive uses and losses of toxic heavy metals in the United States, 1880-1980. In: R.U. Ayres and U.E. Simonis (eds.). *Industrial Metabolism, Restructuring for Sustainable Development*, United Nations University Press, Tokyo, Japan.

Ayres, R.U. and U.E. Simonis (eds.) (1994). *Industrial Metabolism, Restructuring for Sustainable Development*, United Nations University Press, Tokyo, Japan.

Ayres, R.U. (1995). Thermodynamics and process analysis for future economic scenarios, *Environmental and Resource Economics*, Vol. 6(3):207-230.

Ayres, R.U. and L.W. Ayres (1996). *Industrial Ecology - Towards Closing the Materials Cycle*, Edward Elgar, Cheltenham UK and Brookfield US.

Ayres, R.U. (1998). Industrial metabolism: work in progress. In: J.C.J.M. van den Bergh and M.W. Hofkes (eds.). *Theory and Implementation of Economic Models for Sustainable Development*, Kluwer Academic Publishers, Dordrecht, forthcoming.

Barde, J.-P. (1995). Environmental policy and policy instruments. In: H. Folmer, H. Landis Gabel and H. Opschoor (eds.). *Principles of Environmental and Resource Economics - a Guide for Students and Decision Makers*, Edward Elgar, Aldershot UK and Brookfield US.

Barro, R.J. and X. Sala-i-Martin (1995). *Economic Growth*, MacGraw Hill, New York.

Baumol, W.J. and W.E. Oates (1988). *The Theory of Environmental Policy*, second edition, Cambridge University Press, Cambridge.

Beers, C.P. van and J.C.J.M. van den Bergh (1995). Internationale handel, milieu en de GATT/WHO (International trade, environment and the GATT/WTO), *Milieu*, Vol. 10:56-64 (in Dutch).

Bekkers, R.J.M. and H.A.J. Mulder (1990). Secundair aluminium: eerste keus?! (Secondary aluminium: first choice?!), IVEM, Groningen.

Berg N.W. van den, C.E. Dutilh and G. Huppes (eds.) (1995). Beginning LCA; a guide into environmental Life-Cycle Assessment, Report 9453 of Dutch National Reuse of Waste Research Programme, Rotterdam.

Berg, D. van den (1994). Materials in the picture, dynamic life cycle analysis of televisions, IVEM-doctoraalverslag nr. 4, Groningen (in Dutch).

Bergbäck, B., S. Anderberg and U. Lohm (1992). Lead load: historical patterns of lead use in Sweden, *Ambio*, Vol. 21:159-165.

Bergh, J.C.J.M. van den and P. Nijkamp (1994). Dynamic macro modelling and materials balance: Economic–environmental integration for sustainable development, *Economic Modelling*, Vol. 11:283–307.

Bergh, J.C.J.M. van den (1996). *Ecological economics and sustainable development: theory, methods and applications*, Edward Elgar, Cheltenham UK and Aldershot US.

Bergman, L. (1990). The development of computable general equilibrium modeling. In: L. Bergman, D.W. Jorgenson and E. Zalai (eds.). *General Equilibrium Modeling and Economic Policy Analysis*, Blackwell, Oxford UK and Cambridge US.

Bergman, L., D.W. Jorgenson and E. Zalai (eds.) (1990). *General Equilibrium Modeling and Economic Policy Analysis*, Blackwell, Oxford UK and Cambridge US.

Bergman, L. (1995). Environment-economy interactions in a computable general equilibrium model: a case study of Sweden. In: P.-O. Johansson, B. Kriström and K.-G. Mäler (eds.). *Current Issues in Environmental Economics*, Manchester University Press, Manchester, UK.

Bertolini, G. (1994). Wastepaper cycle management: incentives and product chain pressure point or 'leverage point' analysis. In: J.B. Opschoor and R.K. Turner (eds.). *Economic Incentives and Environmental Policies*, Kluwer Academic Publishers, Dordrecht, pp. 229-249.

Beukering, P. van and A. Duraiappah (1996). The economic and environmental impacts of the waste paper trade and recycling in India: a material balance approach, CREED Working Paper Series no. 10, Amsterdam.

Beukering, P. (1997). International trade and recycling in developing countries: waste paper in India, *Warmer Bulletin*, Vol. 52:8-9.

Bianciardi, C., E. Tiezzi and S. Ulgiati (1993). Complete recycling of matter in the frameworks of physics, biology and ecological economics, *Ecological Economics*, Vol. 8:1-5.

Blanchard, O.J. and S. Fischer (1989). *Lectures on Macroeconomics*, MIT Press, Cambridge MA.

Boelens, J. and X. Olsthoorn (1998). Software for material flow analysis. In: P. Vellinga, J. Gupta and F. Berkhout (eds.). *Substantial Sustainability*, Kluwer Academic Publishers, Dordrecht, forthcoming.

Bohm, P. (1981). *Deposit-refund Systems: Theory and Applications to Environmental, Conservation, and Consumer Policy*, Resources for the Future, Washington.

Bohm, P. (1997). Environmental taxation and the double dividend: fact or fallacy?. In: T. O'Riordan (ed.). *Ecotaxation*, Earthscan Publications, London.

Boo, B. de, P. Bosch, C. Gorter and S. Keuning (1991). An environmental module and the complete system of national accounts, Netherlands Central Bureau of Statistics, Nota nr. 154-91-STD.E8, Voorburg, the Netherlands.

Boons, F. (1995). *Produkten in Ketens - an institutionele analyse van de substitutie van PVC-*

leidingsystemen en melkverpakkingen (Products in Chains - an institutional analysis of the substitution of PVC-piping and milk packaging), Ph.D. thesis, Tilburg University Press.

Boulding, K.E. (1978). *Ecodynamics - a new Theory of Societal Evolution*, Sage Publications, Beverly Hills CA and London UK.

Boulding, K.E. (1981). *Evolutionary Economics*, Sage Publications, Beverly Hills CA.

Bovenberg, A.L. and R.A. de Mooij (1994). Environmental levies and distortionary taxation, *American Economic Review*, Vol. 94:1085-1089.

Bovenberg, A.L. and S. Smulders (1995). Environmental quality and pollution-augmenting technological change in a two-sector endogenous growth model, *Journal of Public Economics*, Vol. 57:369-391.

Bovenberg, A.L. and S. Smulders (1996). Transitional impacts of environmental policy in an endogenous growth model, *International Economic Review*, forthcoming.

Bovenberg, A.L. and R.A. de Mooij (1996). Environmental tax reform and endogenous growth, *Journal of Public Economics*, forthcoming.

Brezet, H. (1994). *Van prototype to standaard; De diffusie van energiebesparende technologie* (From prototype to standard; the diffusion of energy-saving technology), Uitgeverij Denhatex, Rotterdam (in Dutch).

Bringezu. S. (1993). Resource intensity analysis - a screening step for LCA, *Fresenius Environmental Bulletin*, Vol. 2, No. 8.

Brinkley, A. (1994). Ecoprofile studies of fabrication methods for IBM computers: sheet metal computer cover, Proceedings of IEEE International Symposium on Electronics and the Environment, pp. 299-304.

Brisson, I. (1993). Packaging waste and the environment: economics and policy, *Resources, Conservation and Recycling*, Vol. 8:183-292.

Brunner, P.H., H. Daxbeck and P. Baccini (1994). Industrial metabolism at the regional and local level: a case-study on the Swiss region. In: R.U. Ayres and U.E. Simonis (eds.) (1994). *Industrial Metabolism, Restructuring for Sustainable Development*, United Nations University Press, Tokyo, Japan.

Bruyn, S.M. de, J.C.J.M. van den Bergh and J.B. Opschoor (1997). Structural change and dematerialisation. In: J. van der Straaten and J.C.J.M. van den Bergh (eds.). *Economics and Ecosystems in Change*, Edward Elgar, Cheltenham UK and Aldershot US.

Bruyn, S.M. and J.B. Opschoor (1997). Developments in the throughput-income relationship: theoretical and empirical observations, *Ecological Economics*, Vol. 20:255-268.

Carraro, C., Y. Katsoulacos and A. Xepapadaeas (eds.) (1996). *Environmental Policy and Market Structure*, Kluwer Academic Publishers, Dordrecht.

CBS (1991). Tax incidence in the Netherlands: accounting and simulations, CBS statistische onderzoekingen M42, SDU Publishers, The Hague.

Chiras, D.D. (1994). *Environmental Science - Action for a Sustainable Future*, 4th edition, The Benjamin/Cummings, California.

Clark, N., F. Perez-Trejo and P. Allen (1995). *Evolutionary Dynamics and Sustainable Development: A Systems Approach*, Edward Elgar, Aldershot UK and Vermont US.

Coase, R.H. (1960). The problem of social cost, *Journal of Law and Economics*, Vol. 3:1-44.

Common, M. (1995). *Sustainability and Policy: Limits to Economics*, Cambridge University Press, Sydney.

Conrad, K. and M. Schröder (1993). Choosing environmental policy instruments using general equilibrium models, *Journal of Policy Modeling*, Vol. 15:521-545.

Conn, W.D. (1977). Consumer product life extension in the context of materials and energy flows. In: Pearce, D.W. and I. Walter (eds.) (1977). *Social and Economic Dimensions of Recycling*, New York University Press, New York.

Copeland, B.R. (1991). International trade in waste products in the presence of illegal disposal, *Journal of Environmental Economics and Management*, Vol. 20:143-162.

Cornielje, O. (1990). *Rationing and Capital Mobility in Applied General Equilibrium Models*, VU University Press, Amsterdam.

Costanza, R., H.E. Daly and J.A. Bartholomew (1990). Goals, agenda and policy recommendations for ecological economics. In: R. Costanza (ed.). *Ecological Economics: the Science and Management of Sustainability*, Columbia University Press, New York.

Cropper, M.L. and W.E. Oates (1992). Environmental economics: a survey, *Journal of Economic Literature*, Vol. 30:675-740.

Cumberland, J.H. (1966). *A Regional Interindustry Model for Analysis of Development Objectives*, Regional Science Association Papers, Vol. 17:65-94.

Dales, J.H. (1968). *Pollution, Property and Prices*, University of Toronto Press, Toronto.

Daly, H. (1968). On economics as a life science, *Journal of Political Economy*, Vol. 76: 392-406.

Daly, H.E. (1991). *Steady-state Economics*, 2nd edition, Island Press, Washington D.C.

Daly, H.E. and J.B. Cobb (1994). *For the Common Good*, 2nd edition, Beacon Press, Boston.

Daly, H.E. (1996). *Beyond Growth - the Economics of Sustainable Development*, Beacon Press, Boston, US.

Deacon, R.T. (1993). Taxation, depletion, and welfare: a simulation study of the U.S. petroleum resource, *Journal of Environmental Economics and Management*, Vol. 24:159-187.

Deacon, R.T. (1995). Assessing the relationship between government policy and deforestation, *Journal of Environmental Economics and Management*, Vol. 28:1-18.

Dellink, R.B. and H.M.A. Jansen (1995). Socio-economic aspects of the greenhouse effect: Applied general equilibrium model, W-95/26, Institute of Environmental Studies, Amsterdam, the Netherlands.

Dellink, R., M. Bennis and H. Verbruggen (1996). Sustainable economic structures: scenarios for sustainability in the Netherlands, W96/27, Institute for Environmental Studies, Amsterdam.

Destais, G. (1996). Economic effects of environmental policies and constraints: what can we learn from computable general equilibrium models? In: S. Faucheux, D. Pearce and J. Proops (eds.). *Models of Sustainable Development*, Edward Elgar, Cheltenham UK and Brookfield US.

Dinan, T.M. (1993). Economic efficiency effects of alternative policies for reducing waste disposal, *Journal of Environmental Economics and Management*, Vol. 25:242-256.

Dinwiddy, C.L. and F.J. Teal (1988). *The Two-sector General Equilibrium Model - a new Approach*, Philip Allan Publishers, Oxford.

Dosi, G., C. Freeman, R. Nelson, G. Silverberg and L. Soete (1988) (eds.). *Technical Change and Economic Theory*. Pinter Publishers, London.

Duchin, F. and D. Szyld (1985). A dynamic input-output model with assured positive output, *Metroeconomica*, Vol. 37:269-282.

Duchin, F. (1988). Analysing structural change in the economy. In: M. Ciaschini (ed.). *Input-output Analysis: Current Developments*, Chapman and Hall, London.

Duchin, F. (1992). Industrial input-output analysis: implications for industrial ecology, *Proceedings of the National Academy of Sciences*, Vol. 89:851-855.

Duchin, F. and G.M. Lange with K. Thornstad and A. Idenburg (1994). *The Future of the Environment: Ecological Economics and Technological Change*, Oxford University Press, New York and Oxford.

Dung, T.H. (1992). Consumption, production, and technological progress: a unified entropic approach, *Ecological Economics*, Vol. 6:195-210.

Eggert, R.G. (1986). Changing patterns of materials use in the U.S. automobile industry, *Materials and Society*, Vol. 10:405-431.

Eggert, R.G. (1990). The passenger car industry: faithful to steel. In: J.E. Tilton (ed.). *World Metal Demand - Trends and Prospects*, Resources for the Future, Washington, D.C.

Eismont, O. (1994). Economic growth with environmental damage and technical progress, *Environmental and Resource Economics*, Vol. 4:241-249.

Ekins, P. (1997). The Kuznets curve for the environment and economic growth: examining the evidence, *Environment and Planning A*, Vol. 29:805-830.

Elburg, G.M. van, P.F. Kroesen, H.C. Moll and W. Biesiot (1992). Verspilling van materialen: cascades en materiaalstromen (Squandering of materials: cascades and material flows), VROM report no 1992/13, The Hague, the Netherlands (in Dutch).

European Commission (1992). The climate challenge: economic aspects of the community's strategy for limiting CO_2 emissions, European Economy, No. 51, Brussels.

European Commission (1993). Taxation, employment and environment: fiscal reform for reducing unemployment, Brussels.

Eyerer, P., 1993. Ganzheitliche Bilanzierung am Beispiel von Kraftfahrzeugkomponenten, VDI-Berichte 1093, VDI-Verlag, Düsseldorf, pp. 73-100.

Faber, M., H. Niemes and G. Stephan (1989). *Entropy, Environment and Resources: an Essay in Physico-economics*, Springer-Verlag, Berlin.

Faber, M. and J.L.R. Proops (1992). *Evolution, Time, Production and the Environment*, 2nd edition, Springer-Verlag, Berlin.

Fankhauser, S. and D. McCoy (1995). Modelling the economic consequences of environmental policies. In: H. Folmer, H. Landis Gabel and H. Opschoor (eds.). *Principles of Environmental and Resource Economics*, Edward Elgar, Aldershot UK and Brookfield US.

Färe, R., S. Grosskopf and D. Tyteca (1996). An activity analysis model of the environmental performance of firms - application to fossil-fuel-fired electric utilities, *Ecological Economics*, Vol. 18: 161-175.

Forsund, F.R. (1985). Input-output models, national economic models, and the environment. In: Kneese, A.V. and J.L. Sweeney (eds.). *Handbook of Natural Resource and Energy Economics*, Volume I, North-Holland, Amsterdam.

Fraanje, P., D. Anink and F. de Haas (1992). Vernieuwbare grondstoffen voor de bouw (Renewable materials for the construction sector), Woon Energie, Gouda (in Dutch).

Fraanje, P. and E. Verkuijlen (1996). Balansen van non-ferrometalen in de Nederlandse woningbouw in 1990 (Balances of non-ferrous metals in the Dutch building sector in 1990), IVAM onderzoeksreeks 76, Amsterdam (in Dutch).

Fraanje, P.J. (1997a). Cascading of pinewood, *Resources, Conservation and Recycling*, Vol. 19:21-28.

Fraanje, P.J. (1997b). Cascading of renewable resources hemp and reed, *Industrial Crops and Products*, Vol. 6 (nos. 3,4):201-212.

Freeman III, A.M. (1993). *The Measurement of Environmental and Resource Values: Theory and Methods*, Resources for the Future, Baltimore, US.

Freeman, C. and C. Perez (1988). Structural crises and adjustment, business cycles and investment behaviour. In: G. Dosi, C. Freeman, R. Nelson, G. Silverberg and L. Soete (eds.) (1988). *Technical Change and Economic Theory*, Pinter Publishers, London and New York.

Fresenius Environmental Bulletin (1993), Vol. 2, No. 8.

Frischknecht R., P. Hofstetter, I. Knoepfel, R. Dones and E. Zollinger (eds.) (1994). Oekoinventare für Energiesysteme, Bundesamt für Energiewirtschaft, Bern.

Frosch, R.A. and N.E. Gallopoulos (1990). Toward an industrial ecology, General Motors Research Labs, No. GMR-6959, Warren MI, USA.

Fullerton, D. and T.C. Kinnaman (1995). Garbage, recycling, and illicit burning or dumping, *Journal of Environmental Economics and Management*, Vol. 29:78-91.

Georgescu-Roegen, N. (1971). *The Entropy Law and the Economic Process*, Harvard University Press, Cambridge MA.

Georgescu-Roegen, N. (1976). *Energy and Economic Myths: Institutional and Economic Essays*, Pergamon Press, New York.

Gilbert, A.J. and J.F. Feenstra (1992). *An indicator of sustainable development - diffusion of cadmium*, Instituut voor Milieuvraagstukken, Amsterdam.

Gilbert, A.J. and J.F. Feenstra (1994). A sustainability indicator for the Dutch environmental policy theme 'diffusion': cadmium accumulation in soil, *Ecological economics*, Vol. 9:253:265.

Ginley, D.M. (1994). Material flows in the transport industry, an example of industrial metabolism, *Resources Policy*, Vol. 3:169-181.

Ginsburg, V. and J. Waelbroeck (1981). *Activity Analysis and General Equilibrium Modelling*, North Holland, Amsterdam.

Ginsburg, V. and J. Waelbroeck (1984). Planning models and general equilibrium activity analysis.

In: H.E. Scarf and J.B. Shoven (eds.). *Applied General Equilibrium Analysis*, Cambridge University Press.

Gjostein, N.A. (1986). Automotive materials usage trends, *Materials and Society*, Vol. 10:369-404.

Gorter, J. (1994). Zinkbalans voor Nederland 1990. Deel 1: de economische stromen (Balance of zinc for the Netherlands in 1990. Part 1: the economic flows). In: Kwartaalbericht Milieustatistieken CBS, Den Haag, 1994 (in Dutch).

Gould, S.J. (1997). Darwinian Fundamentalism, *The New York Review*, 12 June, pp. 34-37.

Gowdy, J. (1994). *Coevolutionary Economics: the Economy, Society and the Environment*, Springer-Verlag, Berlin.

Gradus, R. and S. Smulders (1993) The trade-off between environmental care and long term growth - pollution in three prototype growth models, *Journal of Economics*, Vol. 58:25-51.

Graedel, T.E. and B.R. Allenby (1995). *Industrial Ecology*, Prentice Hall, New Jersey.

Gross, L.S. and E.C.H. Veendorp (1990). Growth with exhaustible resources and a materials balance production function, *Natural Resource Modeling*, Vol. 4:77-94.

Guinée, J.B., P.A.A. Mulder, R. Huele, G. Huppes and L. van Oers (1991). SIMAPRO, a system for the integral environmental analysis for products, Centre of Environmental Science, Leiden, the Netherlands.

Guinée, J.B., J.G.M. Kortman, P.A.A. Mulder and E.W. Lindeijer (1992). SIMAKOZA, a system for the integral environmental analysis of window frames, Centre of Environmental Science-report no. 82, Leiden, the Netherlands.

Guinée, J.B., H.A. Udo de Haes and G. Huppes (1993a). Quantitative life cycle assessment of products, Part 1: Goal definition and inventory, *Journal of Cleaner Production*, Vol. 1(1):3-13.

Guinée, J.B., H.A. Udo de Haes and G. Huppes (1993b). Quantitative life cycle assessment of products, Part 2: Classification, valuation and improvement analysis, *Journal of Cleaner Production*, Vol. 1(2):82-91.

Guinée, J.B. (1995). *Development of a methodology for the environmental life-cycle assessment of products - with a case study on margarines*, Ph.D. thesis, Leiden.

Guinée, J.B., J.C.J.M. van den Bergh, J. Boelens, P.J. Fraanje, G. Huppes, P.P.A.A.H. Kandelaars, Th.M. Lexmond, S.W. Moolenaar, A.A. Olsthoorn, H.A. Udo de Haes, E. Verkuijlen and E. van der Voet (1999). Evaluation of risks of metal flows and accumulation in economy and environment, *Ecological Economics*, forthcoming.

Haan, de M. and S.J. Keuning (1994). Een uitbreiding van de Nationale Rekeningen met een systeem van milieurekeningen (An extension of the National Accounts with a system of environmental accounts), Centraal Bureau voor de Statistiek, BPA-mr. 7501-94-H.E8, Voorburg, the Netherlands (in Dutch).

Hafkamp, W. (1991). Three decades of environmental-economic modelling. In: F. Dietz, F. van der Ploeg and J. van der Straaten (eds.). *Environmental Policy and the Economy*, Elsevier Science Publishers, Netherlands.

Hamilton, C. (1997). The sustainability of logging in Indonesia's tropical forests: a dynamic input-output analysis, *Ecological Economics*, Vol. 21:183-195.

Hamond, M.J., S.J. DeCano, P. Duxbury, A.H. Sanstad and C.H. Stinson (1997). Tax waste, not work, Redefining Progress, San Francisco CA.

Han, X. and T.K. Lakshmanan (1994). Structural changes and energy consumption in the Japanese economy 1975-85: an input-output analysis, *The Energy Journal*, Vol. 15(3):165-188.

Hanley, N. and C.L. Spash (1993). *Cost-benefit Analysis and the Environment*, Edward Elgar, Cheltenham, UK and Brookfield, US.

Hannon, B. and M. Ruth (1994). *Dynamic Modeling*, Springer-Verlag, Berlin.

Hartog, H. den and R.J.M. Maas (1990). Een duurzame economische ontwikkeling: macro-economische aspecten van een prioriteit voor het milieu (A sustainable economic development: macroeconomic aspects of a priority for the environment). In: P. Nijkamp and H. Verbruggen (eds.). *Het Nederlandse Milieu in de Europese Ruimte* (The Dutch environment in the European space), Stenfert Kroese Uitgevers, Leiden, the Netherlands (in Dutch).

Heijnis, R. (1990). *Prisma van het Milieu - ca. 2000 Begrippen van A tot Z Verklaard*, Het Spectrum

BV, Utrecht (in Dutch).

Heijungs, R. (1997). *Economic Drama and the Environmental Stage - Formal Derivation of Algorithmic Tools for Environmental Analysis and Decision-Support from a Unified Epistemological Principle*, Ph.d. thesis, Centrum voor Milieukunde, Leiden, the Netherlands.

Heintz, R. and H. Verbruggen (1997). Meer groei en toch een schoner milieu? De groene Kuznets curve (Do economic growth and a clean environment go hand in hand? The environmental Kuznets curve), *Milieu*, Vol. 12:2-9 (in Dutch).

Hendrix, C. and C. Martens (1990). Het milieuvriendelijkste kozijn. Een passend raamwerk binnen het normkostensysteem? (The most environmentally-friendly window frame. A suitable framework in the standard cost system), Publikatiefonds milieukunde Nieuw Rollecate nr. 8, Deventer (in Dutch).

Henry, E.W. (1994). A capacity growth input-output model, *Energy Economics*, Vol. 16(3):193-203.

Herman, R., S.A. Ardekani and J.H. Ausubel (1989). Dematerialization. In: J.H. Ausubel and H.E. Sladovich (eds.). *Technology and Environment*, National Academic Press, Washington.

Hertel, T.W. (1988). General equilibrium incidence of natural resource subsidies: the three factor case, *Journal of Environmental Economics and Management*, Vol. 15:206-223.

Hinterberger, F., F. Luks and F. Schmidt-Bleek (1995). What is "Natural Capital"?, Wuppertal Papers nr. 29, Wuppertal, Germany.

Hoefnagels, F.E.T., J.G.M. Kortman and E.W. Lindeijer (1992). Minimalisering van de milieubelasting van buitenkozijnen in de woningbouw (Minimization of the environmental burden of window frames in the construction of houses), IVAM onderzoeksreeks nr. 54, Amsterdam (in Dutch).

Hoel, M. (1978). Resource extraction and recycling with environmental costs, *Journal of Environmental Economics and Management*, Vol. 5:220-235.

Hoevenagel, R. (1994). *The Contingent Valuation Method: Scope and Validity*, Ph.D. thesis, Institute for Environmental Studies, Amsterdam.

Holling, C.S., D.W. Schindler, B.W. Walker and J. Roughgarden (1995). Biodiversity in the functioning of ecosystems: an ecological synthesis. In: C. Perrings, C.S. Holling, K.-G. Mäler, B.-O. Jansson and C. Folke (eds.). *Biodiversity Loss - Economic and Ecological Issues*, Cambridge University Press, Cambrigde.

Holmes, M.J. (1990). Materials substitution in the UK window industry, 1983-87, *Resources Policy*, Vol. 16:128-143.

Hond, F. den (1996). *In Search of a Useful Theory of Environmental Strategy: a Case Study on the Recycling of End-of-Life Vehicles from the Capabilities Perspective*, Ph.d. thesis, Vrije Universiteit, Amsterdam.

Hotelling, H. (1931). The economics of exhaustible resources, *Journal of Political Economy*, Vol. 39: 137-175.

Howarth, R.B. and R.B. Norgaard (1993). Intergenerational transfers and the social discount rate, *Environmental and Resource Economics*, Vol. 3:337-358.

Howarth, R.B. and R.B. Norgaard (1995). Intergenerational choices under global environmental change. In: D.W. Bromley (ed.). *The Handbook of Environmental Economics*, Blackwell, Oxford UK and Cambridge US.

Huang, G.H., W.P. Anderson and B.W. Baetz (1993). Environmental input-output analysis and its application to regional solid-waste management planning, *Journal of Environmental Management*, Vol. 42:63-79.

Huang, C.-H. and D. Cai (1994). Constant-returns endogenous growth with pollution control, *Environmental and Resource Economics*, Vol. 4:383-400.

Hueting, R. (1991). Correcting national income for environmental losses: a practical solution for a theoretical dilemma. In: R. Costanza (ed.). *Ecological Economics*, Columbia University Press, New York.

Huppes, G. (1993). *Macro-Environmental Policy: Principles and Design*, Elsevier Science Publishers BV, Amsterdam.

Husar, R.B. (1994). Sulphur and nitrogen emission trends for the United States: an application of the

materials flow approach. In: R.U. Ayres and U.E. Simonis (eds.). *Industrial Metabolism, Restructuring for Sustainable Development*, United Nations University Press, Tokyo, Japan.
Idenburg, A.M. and A.E. Steenge (1991). Environmental policy in single-product and joint production input-output models. In: F. Dietz, F. van der Ploeg and J. van der Straaten (eds.). *Environmental Policy and the Economy*, Elsevier Science Publishers, Dordrecht.
Idenburg, A.M. (1993). *Gearing Production Models to Ecological Economic Analysis: a Case Study, within the Input-Output Framework, of Fuels for Road Transport*, University Twente, Enschede, the Netherlands.
Isard, W. *et al.* (1971). *Ecologic-Economic Analysis for Regional Planning*, The Free Press, New York.
ISO (1995). ISO Environmental Management Standardization Efforts, NISTIR 5638, Brussels.
James, D.E. (1985). Environmental economics, industrial process models, and regional residuals management models. In: A.V. Kneese and J.L. Sweeney (eds.). *Handbook of Natural Resource and Energy Economics*, Volume I, North-Holland, Amsterdam.
James, D.E., H.M.A. Jansen and J.B. Opschoor (1978). *Economic Approaches to Environmental Problems*, Elsevier, Amsterdam, the Netherlands.
Jansen, M.H. and A.J.D. Lambert (1996). Environmental information systems based on mass and energy flows for companies in supply chains, *Milieu*, Vol. 11:123-129 (in Dutch).
Janssen, R. (1992). *Multiobjective Decision Support for Environmental Management*, Kluwer Academic Publishers, Dordrecht.
Jolly, J.H. (1992). Materials flow of zinc in the United States 1850-1990, Bureau of Mines OFR# 72-92.
Jorgenson, D. and P. Wilcoxen (1993). Reducing U.S. carbon dioxide emissions: an assessment of different instruments, *Journal of Policy Modeling*, Vol. 15:491-520.
Kandelaars, P.P.A.A.H. and J.C.J.M. van den Bergh (1996a). Materials-product chains: theory and an application to zinc and pvc gutters, *Environmental and Resource Economics*, Vol. 8:97-118.
Kandelaars, P.P.A.A.H. and J.C.J.M. van den Bergh (1996b). Materials-product chain analysis of window frames, Tinbergen Institute Discussion Paper 5-96-12, Amsterdam.
Kandelaars, P.P.A.A.H. and J.C.J.M. van den Bergh (1996c). Economische beleidsanalyse van materiaal-product ketens toegepast op de zinkproblematiek en dakgoten, *Milieu*, Vol. 4:178-186.
Kandelaars, P.P.A.A.H., J.B. Opschoor and J.C.J.M. van den Bergh (1996). A dynamic simulation model for materials-product chains: An application to gutters. *Journal of Environmental Systems*, Vol. 24 (4):345-371.
Kandelaars, P.P.A.A.H., M.H. Jansen and A.J.D. Lambert (1996). Survey of methods of material and product flow analysis, Tinbergen Institute Discussion Paper 96-124/5, Amsterdam.
Kandelaars, P.P.A.A.H. and J.C.J.M. van den Bergh (1997a). Dynamic analysis of materials-product chains: An application to window frames, *Ecological Economics*, Vol. 22:41-61.
Kandelaars, P.P.A.A.H. and J.C.J.M. van den Bergh (1997b). General equilibrium analysis of economic instruments in material-product chains with material balance, recycling and waste treatment, Tinbergen Institute Discussion Paper 97-110/3, Amsterdam.
Kandelaars, P.P.A.A.H. and J.C.J.M. van den Bergh (1998). Integrated chain analysis of material and product flows under alternative environmental policy packages. In: S. Dwyer, U. Ganslasser and M. O'Connor (eds.). *Ecology, Society, Economy: Life Sciences Dimensions*, Filander Press, Germany, forthcoming.
Kandelaars, P.P.A.A.H. and J.D. van Dam (1998). An analysis of variables influencing the material composition of automobiles, *Resources, Conservation and Recycling*, Vol. 24:323-333.
Kandelaars, P.P.A.A.H. and R.B. Dellink (1999). Economic effects of material policies: combining an applied equilibrium model with physical flows, *Journal of Policy Modeling*, forthcoming.
Keeler, A.G. and M. Renkow (1994). Haul trash or haul ash: energy recovery as a component of local solid waste management, *Journal of Environmental Economics and Management*, Vol. 27:205-217.
Keller, W.J. (1980). *Tax Incidence: A General Equilibrium Approach*, North-Holland, Amsterdam.
Kemp, R. (1995). *Environmental Policy and Technical Change: a Comparison of the Technological*

Impact of Policy Instruments, Universitaire Pers, Maastricht.

Kneese, A.V., R.U. Ayres and R.C. d'Arge (1970). *Economics and the Environment: A Materials Balance Approach*. Johns Hopkins Press, Baltimore.

Konijn, P.J.A. (1995). Description of materials flows in the economy, *Milieu*, Vol. 10(3):121-127 (in Dutch).

Leontief, W. (1966). *Input-Output Economics*, Oxford University Press, New York.

Leontief, W. (1970). Environmental repercussions and the economic structure: an input-output approach, *Review of Economics and Statistics*, Vol. 52:262-271.

Leontief, W. and D. Ford (1972). Air pollution and the economic structure: empirical results of input-output computations. In: A. Brody and A.P. Carter (eds.). *Input-Output Techniques*, North-Holland, Amsterdam.

Leontief, W., A. Carter and P. Petri (1977). *Future of the World Economy*, Oxford University Press, New York.

Leontief, W. (1986). Input-output analysis. In: M.B. Bever (ed.). *Encyclopedia of Materials, Science and Engineering*, Pergamon, Oxford.

Lévêque, F. and A. Nadaï (1995). A firm's involvement in the policy-making process. In: H. Folmer, H. Landis Gabel and H. Opschoor (eds.). *Principles of Environmental and Resource Economics - a Guide for Students and Decision Makers*, Edward Elgar, Aldershot UK and Brookfield US.

Lindeijer, E., O. Mekel, G. Huppes and R. Huele (1990). Milieu-effecten van kozijnen (Environmental effects of window frames); Interim rapportage fase 1, Centrum voor Milieukunde, Leiden (in Dutch).

Lines, M. (1995). Dynamics and uncertainty. In: H. Folmer, H. Landis Gabel and H. Opschoor (eds.). *Principles of Environmental and Resource Economics*, Edward Elgar, Aldershot UK and Brookfield US.

Lintott, J. (1996). Survey of environmental accounting: useful to whom and for what?, *Ecological Economics*, Vol. 16.:179-190.

Löfgren, K.-G. (1995). Markets and externalities. In: H. Folmer, H. Landis Gabel and H. Opschoor (eds.) (1995). *Principles of Environmental and Resource Economics - a Guide for Students and Decision Makers*, Edward Elgar, Aldershot UK and Brookfield US.

Lohm, U., S. Anderberg and B. Bergbäck (1994). Industrial metabolism at the national level: a case-study on chromium and lead pollution in Sweden, 1880-1980. In: R.U. Ayres and U.E. Simonis (eds.). *Industrial Metabolism, Restructuring for Sustainable Development*, United Nations University Press, Tokyo, Japan.

Luptácik, M. and B. Böhm (1994). An environmental input-output model with multiple criteria, *Annals of Operations Research*, Vol. 54:119-127.

Lusky, R. (1975). Optimal taxation policies for conservation and recycling, *Journal of Economic Theory*, Vol. 11:315-328.

Lusky, R. (1976). A model of recycling and pollution control, *Canadian Journal of Economics*, Vol. 9:91-101.

Manne, A. (1977). General equilibrium with activity analysis. In: C. Hitch (ed.). *Modeling Energy-Economy Interactions: Five Approaches*, Resources for the Futures, Washington DC.

Matthijssen, A.J.X.M. and P.J. Meijer (1992). Produktie van primair zink (Production of primary zinc), Samenwerkingsproject Procesbeschrijvingen Industrie Nederland, RIVM-rapportnummer 73601113 (in Dutch).

Max, W. and D.E. Lehman (1988). A behavioral model of timber supply, *Journal of Environmental Economics and Management*, Vol. 15:71-86.

McClain, K.T. (1995). Recycling programs. In: Bromley D.W. (ed.). *The Handbook of Environmental Economics*, Blackwell, Oxford, pp. 223-239.

McNicoll, I.H. and D. Blackmore (1993). A pilot-study on the construction of a Scottish environmental input-output system, Report to Scottish enterprise, Department of Economics, University of Strathclyde, Glasgow, Scotland.

Midmore, P. (1993). Input-output forecasting of regional agricultural policy impacts, *Journal of*

Agricultural Economics, Vol. 44(2):284-300.

Miernyk, W.H. and J.T. Sears (1974). *Air Pollution Abatement and Regional Economic Development*, Lexington Books, D.D. Heath and Co., Lexington MA.

Miller, R.E. and P.D. Blair (1985). *Input-Output Analysis - Foundations and Extensions*, Prentice-Hall, Englewood Cliffs NJ, USA.

Moll, H.C. (1993). *Energy Counts and Materials Matter in Models for Sustainable Development*, STYX publications, Groningen.

Moolenaar, S.W., Th.M. Lexmond and S.E.A.T.M. van der Zee (1997). Calculating heavy metal accumulation in soil: A comparison of methods illustrated by a case-study on compost application, *Agriculture, Ecosystems and Environment*, forthcoming.

Moolenaar, S.W. (1998). *Sustainable management of heavy metals in agro-ecosystems*, Ph.D. thesis, Wageningen.

Morris, G.E. and D.M. Holthausen (1994). The economics of household solid waste generation and disposal, *Journal of Environmental Economics and Management*, Vol. 26:215-234.

Nappi, C. (1986). The food and beverage container industries: change and diversity. In: J.E. Tilton (ed.). *World Metal Demand - Trends and Prospects*, Resources for the Future, Washington, D.C.

Neher, P.A. (1990). *Natural Resource Economics - Conservation and Exploitation*, Cambridge University Press, Cambridge.

Nelson, R.R. and S.G. Winter (1982). *An Evolutionary Theory of Economic Change*, Harvard University Press, Cambridge MA.

Nelson, R.R. (1995). Recent evolutionary theorizing about economic change, *Journal of Economic Literature*, Vol. 33:48-90.

Netherlands Scientific Council for Government Policy (1992). Environmental policy: strategy, instruments and enforcement, Summary of the 41th Report, The Hague.

Nijkerk, F. (1994). Recycling kent ook haar grenzen (Recycling also knows its borders), *Chemisch Magazine*, June 1994, pp. 150-252 (in Dutch).

O'Connor, M. (1993). Entropic irreversibility and uncontrolled technological change in economy and environment, *Journal of Evolutionary Economics*, Vol. 3:285-315.

O'Riordan, T. (ed.) (1997). *Ecotaxation*, Earthscan Publications, London.

Odum, E.P. (1971). *Ecology*, Holt, Reinhard and Winston, London.

OECD (1972). Council recommendation on guiding principles concerning international aspects of environmental policies, 26 May. In: The polluter pays principle, Environment Monograph, Paris.

OECD (1989). Economic instruments for environmental protection, Paris.

OECD (1992). The economic costs of reducing CO_2 emissions, OECD Economic Studies, 19/Winter, Paris, France.

OECD (1994). Applying economic instruments to environmental policies in OECD and dynamic non-member economies, OECD, Paris.

Olsthoorn, A.A., M. van Herwijnen and P.C. Koppert (1991). Ketenbeheer van vliegas met dynamische materiaalbalansen (Chain management of fly ash with dynamic material balances), *Milieu*, Vol. 1:7-11 (in Dutch).

Olsthoorn, A.A. (1991). Sources of persistent micropollutants: analysis with dynamic materials balances. In: J.B. Opschoor and D.W. Pearce (1991). *Persistent Pollutants: Economics and Policy*, Kluwer Academic Publishers, Dordrecht.

Oosterhuis, F., F. Rubik and G. Scholl (1996). *Product Policy in Europe: New Environmental Perspectives*, Kluwer Academic Publishers, Dordrecht.

Opschoor, J.B. and J. van der Straaten (1993). Sustainable development: an institutional approach, *Ecological Economics*, Vol. 7:203-222.

Opschoor, J.B. (1994). Chain management in environmental policy: analytical an evaluative concepts. In: J.B. Opschoor and R.K. Turner (eds.). *Economic Incentives and Environmental Policies*, Kluwer Academic Publishers, Dordrecht.

Opschoor, J.B. and R.K. Turner (eds.) (1994). *Economic Incentives and Environmental Policies*, Kluwer Academic Publishers, Dordrecht.

Opschoor, J.B. (1995). Economische instrumenten in het Nederlandse milieubeleid (Economic instruments in Dutch environmental policy), *Milieu*, Vol. 5:229-236 (in Dutch).

Opschoor, J.B. (1996). Sustainability, Economic Restructuring and Social Change, Institute of Social Studies, the Hague, Inaugural Address, May 23 1996.

Osnowski, R. and F. Rubik (1987). *Produktlinienanalyse - Bedürfnisse, Produkte und ihre Folgen* (Product Analysis - Requirements, Products and their Effects), Kölner Volksblatt Verlag (in German).

Page, T. (1977). *Conservation and Economic Efficiency - an Approach to Materials Policy*, Resources for the Future, John Hopkins University Press, Baltimore.

Pearce, D.W. and I. Walter (eds.) (1977). *Social and Economic Dimensions of Recycling*, New York University Press, New York.

Pearson, P.J.G. (1989). Proactive energy-environment policy strategies: a role for input-output analysis?, *Environment and Planning A*, Vol. 21:1329-1348.

Peet, J. (1992). *Energy and the Ecological Economics of Sustainability*, Island Press, Washington DC, USA.

Perman, R., Y. Ma and J. McGilvray (1996). *Natural Resource and Environmental Economics*, Addison Wesley Longman Limited, Essex, England.

Perrings, C. (1987). *Economy and Environment - a Theoretical Essay on the Interdependence of Economic and Environmental Systems*, Cambridge University Press, Cambridge.

Peterson, S. and B. Richmond (1994). Stella II, High Performance Systems, Hanover, New Hampshire, US.

Pezzey, J. (1992). Sustainability: an interdisciplinary guide, *Environmental Values*, Vol. 1:321-362.

Ploeg, F. van der and C. Withagen (1993). Pollution control and the Ramsey problem, *Environmental and Resource Economics*, Vol. 2:215-230.

Porteous, A. (1996). *Dictionary of Environmental Science and Technology*, second Edition, John Wiley, Chichester, England.

Potier, M. (1977). Summary and Overview. In: D.W. Pearce and I. Walter (eds.). *Social and Economic Dimensions of Recycling*, New York University Press, New York.

Recycling (1994), March/April.

Roberts, M.C. (1992). The material composition of products and new materials, *Resources Policy*, Vol. 18:122-136.

Ruth, M. (1993). *Integrating Economics, Ecology and Thermodynamics*, Kluwer Academic Publishers, Dordrecht.

Ruth, M. (1995a). Thermodynamic constraints on optimal depletion of copper and aluminium in the United States: a dynamic model of substitution and technical change, *Ecological Economics*, Vol. 15:197-213.

Ruth, M. (1995b). Thermodynamic implications for natural resource extraction and technical change in U.S. copper mining, *Environmental and Resource Economics*, Vol. 6:187-206.

Schmidt-Bleek, F. (1993). Fresenius Environmental Bulletin, The International Journal for Rapid Communication and Updating in the field of Biotic and Abiotic Systems, Vol. 2, nr 8.

Schneider, F. (1993). Allocation and recycling: enlarging to the cascade system, CML, Leiden, unpublished paper.

Schuckert, M. (1993). Die ganzheitliche Bilanzierung am Beispiel des Systems Automobil, VDI-Berichte 1093, VDI-Verlag, Düsseldorf, pp. 57-71.

Schumpeter, J.A. (1934). *The Theory of Economic Development*, Harvard University Press, Cambridge MA, USA.

Selden, T.M. and D. Song (1994). Environmental quality and development: is there a Kuznets curve for air pollution emissions?, *Journal of Environmental Economics and Management*, Vol. 27:147-162.

Siebert, H. (1987). *Economics of the Environment - Theory and Policy*, Springer-Verlag, Berlin.

Sirkin, T. and M. ten Houten (1993). Resource cascading and the cascade chain; tool for appropriate and sustainable product design, IVAM Research Report no 71, Amsterdam.

Snowdon, K.G. (1994). Environmental life cycle assessment of face plates used in the telecommunications industry, Proceedings of IEEE International Symposium on Electronics and the

Environment, pp. 293-298.

Socolow, R., C. Andrews, F. Berkhout and V. Thomas (eds.) (1994). *Industrial Ecology and Global Change*, Cambridge University Press, Cambridge.

Solow, R.M. (1974). Intergenerational equity and exhaustible resources, *Review of Economic Studies*, Symposium Economics Exhaustible Resources, pp. 26-46.

Solow, R.M. (1986). On the intergenerational allocation of natural resources, *Scandinavian Journal of Economics*, Vol. 88:141-149.

Starreveld, P.F. and E.C. van Ierland (1994). Recycling of plastics: a material balance optimisation model, *Environmental and resource economics*, Vol. 4:251-264.

Stavins, R.N. (1995). Transactions costs and tradeable permits, *Journal of Environmental Economics and Management*, Vol. 29:33-148.

Stigliani, W.M. and S. Anderberg (1992). Industrial metabolism at the regional level: the Rhine Basin. IIASA Working Paper, Laxenburg, Austria.

Stigliani, W.M. and S. Anderberg (1994). Industrial metabolism at the regional level: The Rhine Basin. In: R.U. Ayres. and U.E. Simonis (eds.). *Industrial Metabolism, Restructuring for Sustainable Development*, United Nations University Press, Tokyo, Japan.

Sullivan, A.M. (1987). Policy options for toxic disposal: laissez-faire, subsidization, and enforcement, *Journal of Environmental Economics and Management*, Vol. 14:58-71.

Tauw Milieu (1994). Milieugerichte levenscyclusanalyse dakgootsystemen (Environmental life-cycle analysis of rain gutter systems), Rapportnummer 3296121, Deventer (in Dutch).

Tietenberg, T.H. (1995). Transferable discharge permits and global warming. In: D.W. Bromley (ed.). *The Handbook of Environmental Economics*, Blackwell, Oxford.

Tietenberg, T.H. (1996). *Environmental and Natural Resource Economics*, 4th edition, HarperCollins College Publishers, New York.

Tilton, J.E. (ed.) (1990). *World Metal Demand - Trends and Prospects*, Resources for the Future, Washington D.C.

Toman, M.A., J. Pezzey and J. Krautkraemer (1995). Neoclassical economic growth theory and 'sustainability'. In: Bromley D.W. (ed.). *The Handbook of Environmental Economics*, Blackwell, Oxford, pp. 139-165.

Tromp, O.-S. (1995a). *Towards Sustainable Quality: a Methodological Principle for Sustainable Management of Materials Use*, Ph.D. thesis, Groningen.

Tromp, O.-S. (1995b). Sustainable use of renewable resources for material purposes - a conceptual approach, United Nations Environment Programme, Amsterdam.

Turner, R.K., R. Grace and D.W. Pearce (1977). The economics of waste paper recycling. In: D.W. Pearce and I. Walter (eds.). *Social and Economic Dimensions of Recycling*, New York University Press, New York.

Turner, R.K. and D. Pearce (1994). The role of economic instruments in solid waste management policy. In: J.B. Opschoor and R.K. Turner (eds.). *Economic Incentives and Environmental Policies*, Kluwer Academic Publishers, Dordrecht.

VCNI (Association of the Dutch Chemical Industry) (1991). Integrated substance chain management, Leidschendam.

Verhoef, E. (1996). *The Economics of Regulating Road Transport*, Edward Elgar, Cheltenham UK and Aldershot US.

Victor, P.A. (1972). *Pollution: Economy and Environment*, George Allen and Unwin Ltd., London.

Voet, E. van der, L. van Egmond, R. Kleijn and G. Huppes (1994a). Cadmium in the European Community: a policy-oriented analysis, *Waste management and research*, No. 12:507-526.

Voet, E. van der (1996). *Substances from Cradle to Grave*, Ph.d. thesis, Optima Druk, Moolenaarsgraaf, the Netherlands.

VROM (Dutch Ministry of Housing, Spatial Planning and Environment) (1989). National environmental policy plan, Den Haag.

VROM (1990). National environmental policy plan-plus, Den Haag, the Netherlands.

VROM (1992). Trendrapport volkshuisvesting 1992 (Report on the trend in the construction of houses 1992), Den Haag.

VROM (1993a). Notitie integraal ketenbeheer (Report on integral chain management), Den Haag, the Netherlands.

VROM (1993b). Nota produkt en milieu (Report product and environment), Den Haag, the Netherlands.

VROM (1993c). National environmental policy plan-2, Den Haag, the Netherlands.

VROM/DGM (1993). Beleidsverklaring milieu-taakstellingen bouw 1995 (Policy statement for the environment in the construction sector 1995), Den Haag.

VROM (1994). Naar een verbeterd kringloopbeheer van aluminium (Towards an improved chain management of aluminium), Rapport nr 1994/13, Den Haag, the Netherlands.

VROM (1995). Milieuaspecten van kozijnen en beglazingsystemen; praktische aanbevelingen voor de bouw (Environmental aspects of window frames and glazing systems; practical recommendations for the construction sector), Den Haag.

Wackernagel, M. and W. Rees (1996). *Our Ecological Footprint: Reducing Human Impact on the Earth*, New Society Publishers, Gabriola Island BC and Philadelphia PA.

WCED (World Commission on Environment and Development) (1987). *Our Common Future*, Oxford University Press, Oxford and New York.

Weaver, P.M., H.L. Gabel, J.M. Bloemhof-Ruwaard and L.N. van Wassenhove (1995). Optimising environmental product life cycles: a case study of the European pulp and paper sector, CMER Working Paper, INSEAD, Fontainebleau.

Weber, C. (1995). Studies on the GER of product classes using input-output analysis, VDI-Berichte 1218, VDI-Verlag, Düsseldorf, pp. 165-189.

Weinstein, M.C. and R.J. Zeckhauser (1974). Use patterns for depletable and recyclable resources, *Review of Economic Studies*, Symposium of Economics of Exhaustible Resources, 67-88.

Werkgroep Vergroening Fiscaal Stelsel (1996) (Committee on the greening of the fiscal system). Tweede rapportage (Second report), the Netherlands (in Dutch).

Wertz, K.L. (1976). Economic factors influencing households' production of waste, *Journal of Environmental Economics and Management*, Vol. 2:263-272.

Wieringa, K. and J.E.C. van den Nieuwenhuijzen (1994). Van monetaire naar fysieke groei - het transformatiemodel (From monetary to physical growth - the transformation model), Rapportnummer 481504002, RIVM, Bilthoven, the Netherlands (in Dutch).

Woodland, A.D. (1980). Direct and indirect trade utility functions, *Review of Economic Studies*, Vol. 47:907-926.

Glossary

Biogeochemical cycle : see Nutrient cycle

Chain management : The management of material flows that result from chains of social and economic activities (VROM, 1993b).

Chain management of M-P chains : The manipulation of M-P chains so as to optimize the environmental impact of these chains, or so as to achieve a certain accepted environmental impact at least social costs (Opschoor, 1994).

Dematerialization : The reduction in the weight of materials used in industrial end products (Herman *et al.*, 1989). From an environmental viewpoint 'the amount of waste generated per unit of industrial products' (Herman *et al.*, 1989)

Double dividend : When environmental taxation leads to positive effects on both the environment and employment.

Ecosystem : The environment of a community of organisms and all the interactions between organisms, and between organisms and their environment (Chiras, 1994)

Ecology : The science that studies relationships of living organisms with their biotic and abiotic environment.

Entropy : A measure of unavailable energy (see Thermodynamics, second law).

Equilibrium model : A model in which agents optimize behaviour, budgets balance and all markets are cleared; A general equilibrium model includes all markets, while a partial equilibrium considers one market while taking the others as exogenous; An applied general equilibrium model is applied to, for example, a country or a region for which the effects of a certain policy on economic groups, including the government, are examined.

Externality : An externality is present when an individual's utility or production function is affected by the behaviour of another individual who does not take into account the effects of his behaviour on the other individual (Baumol and Oates, 1988).

Food chain : A series of organisms, each feeding on the preceding one (Chiras, 1994).

Industrial ecology : A concept similar to 'industrial metabolism' (see below). It may be interpreted as industrial metabolism extended with a human perspective in which, for example, consumption and preferences are included (Duchin, 1992; Graedel and Allenby, 1995)

Industrial ecosystem : A system designed from scratch to imitate nature by utilizing the waste products of each component firm as raw materials (or 'food') for another firm (Ayres and Ayres, 1996).

Industrial metabolism : The set of physico-chemical transformations that convert raw materials (biomass, fuels, minerals, metals) into manufactured products and structures and wastes (Ayres and Simonis, 1994). Industrial metabolism is an approach to modify the use of materials in order to reduce the generation of waste by applying lessons from the natural world (Duchin, 1992).

Input-output analysis : An input-output table describes, originally in monetary units, the mutual exchange of goods and services between different industrial sectors and towards final users.

Integral chain management : The adoption of the chain management approach in all environmental policy making and by all economic agents (VROM, 1993b).

Life-cycle assessment : A method to assess the inputs and outputs of materials and energy and the associated environmental impacts that are directly attributable to the functioning of a product throughout its life cycle (Guinée, 1995).

Life cycle of a product : Refers to the material and energy flows from a product from extraction to final discharge (Guinée, 1995).

Material balance (MB) principle : In a closed system materials cannot be lost, only altered (derived from the first law of thermodynamics). For an open system the MB principle implies that all materials that go into a system either accumulate or leave the system. (Ayres, 1978; Van den Bergh, 1996).

Material flow analysis (MFA) : A method to describe the flow of a specific material in a certain geographic region in a certain time period (Van der Voet, 1996).

Material-product (M-P) chain : A subset of linked material and product flows for a certain service or application (Opschoor, 1994). This refers to a network of economic activities between extraction and waste treatment, connected via flows of materials and products. An M-P chain is a chain of processes in economic and physical dimensions. This combination of physical flows and economic product flows in an M-P chain allows one to include recycling of materials, reuse of products and substitution.

Material-product (M-P) chain analysis (broad definition): An analysis of an economic structure of connected material and product flows.

Material-product (M-P) chain analysis (narrow definition as in this study) : A study on the allocation and economic processes of an M-P chain. M-P chain analysis is used for the study of optimization, system analysis, market equilibrium, market processes, production functions, policy analysis, substitution at different levels, explicit modelling of economic processes, and endogenous behaviour of agents.

Metabolism (biological) : The totality of internal processes of a living organism, i.e. the ingestion of materials (food) and the generation of waste outputs (excretion and exhalation) (Ayres, 1997).

Miniaturization : The use of small quantities of (various) materials in products (Tromp, 1995a).

Nutrient cycle : The flow of nutrients, like carbon and nitrogen, in ecosystems (Odum, 1971).

Primary materials : Newly extracted materials

Recycling (of materials) : Used materials are converted by physical or chemical processes to be used again (Porteous, 1996).

Reuse (of products) : Used products are used again without converting the product physically or chemically.

Secondary materials : Recycled materials

Substance flow analysis (SFA) : see Material flow analysis

Substitution, direct : Substitution between materials.

Substitution, indirect : Substitution between materials and non-materials

Succession : The sequential development of a community of species (Chiras, 1994).

Thermodynamics, first law : The law of conservation of energy states that in a closed system energy cannot be created or destroyed only transformed (Ruth, 1993).

Thermodynamics, second law : This law states that the entropy - a measure of unavailable energy - of an isolated system will not decrease (Ayres 1978; Peet, 1992; Ruth, 1993).

Throughput : The flow beginning with raw material inputs, followed by their conversion into commodities, and finally into waste outputs (Daly, 1996).

Eco-Efficiency in Industry and Science

1. J.E.M. Klostermann and A. Tukker (eds.): *Product Innovation and Eco-efficiency.* Twenty-three Industry Efforts to Reach the Factor 4. 1997 ISBN 0-7923-4761-7
2. K. van Dijken, Y. Prince, T. Wolters, M. Frey, G. Mussati, P. Kalff, O. Hansen, S. Kerndrup, B. Søndergård, E. Lopes Rodrigues and S. Meredith (eds.): *Adoption of Environmental Innovations.* The Dynamics of Innovation as Interplay Between Business Competence, Environmental Orientation and Network Involvement. 1999 ISBN 0-7923-5561-X
3. M. Bartolomeo, M. Bennett, J.J. Bouma, P. Heydkamp, P. James, F. de Walle and T. Wolters: *Eco-Management Accounting.* 1999 ISBN 0-7923-5562-8
4. P.P.A.A.H. Kandelaars: *Economic Models of Material-Product Chains for Environmental Policy Analysis.* 1999 ISBN 0-7923-5794-9

KLUWER ACADEMIC PUBLISHERS – DORDRECHT / BOSTON / LONDON

MIX
Papier aus verantwortungsvollen Quellen
Paper from responsible sources
FSC® C105338

If you have any concerns about our products,
you can contact us on
ProductSafety@springernature.com

In case Publisher is established outside the EU,
the EU authorized representative is:
Springer Nature Customer Service Center GmbH
Europaplatz 3, 69115 Heidelberg, Germany

Printed by Libri Plureos GmbH
in Hamburg, Germany